Cyril Norwood and the Ideal of
Secondary Education

Secondary Education in a Changing World

Series editors: Barry M. Franklin and Gary McCulloch

Published by Palgrave Macmillan:

The Comprehensive Public High School: Historical Perspectives
By Geoffrey Sherington and Craig Campbell (2006)

Cyril Norwood and the Ideal of Secondary Education
By Gary McCulloch (2007)

Cyril Norwood and the Ideal of Secondary Education

Gary McCulloch

CYRIL NORWOOD AND THE IDEAL OF SECONDARY EDUCATION
© Gary McCulloch 2007.
Softcover reprint of the hardcover 1st edition 2007 978-1-4039-6793-0

All rights reserved. No part of this book may be used or reproduced in any manner whatsoever without written permission except in the case of brief quotations embodied in critical articles or reviews.

First published in 2007 by
PALGRAVE MACMILLAN™
175 Fifth Avenue, New York, N.Y. 10010 and
Houndmills, Basingstoke, Hampshire, England RG21 6XS
Companies and representatives throughout the world.

PALGRAVE MACMILLAN is the global academic imprint of the Palgrave Macmillan division of St. Martin's Press, LLC and of Palgrave Macmillan Ltd. Macmillan® is a registered trademark in the United States, United Kingdom and other countries. Palgrave is a registered trademark in the European Union and other countries.

ISBN 978-1-349-53036-6   ISBN 978-0-230-60352-3 (eBook)
DOI 10.1057/978023060352 3

Library of Congress Cataloging-in-Publication Data

McCulloch, Gary.
    Cyril Norwood and the ideal of secondary education / Gary McCulloch.
    p. cm.—(Secondary education in a changing world)
    Includes bibliographical references and index.
    Contents: Introduction: Cyril Norwood and secondary education—Middle class education and the State—The education of Cyril Norwood—The higher education of boys in England—Holding the line?—Marlborough and Harrow—The English tradition of education—The new world of education—The Norwood Report and secondary education—Conclusions: the ideal of secondary education.

    1. Education, Secondary—Great Britain—History—20th century.
    2. Norwood, Cyril, Sir, 1875– I. Title.

LA634.M289 2007
373.41—dc22   2006044677

A catalogue record for this book is available from the British Library.

Design by Newgen Imaging Systems (P) Ltd., Chennai, India.

First edition: March 2007
10 9 8 7 6 5 4 3 2 1

Transferred to Digital Printing 2011

Contents

Preface		vi
Acknowledgments		viii
Chapter 1	Introduction: Cyril Norwood and Secondary Education	1
Chapter 2	Middle Class Education and the State	11
Chapter 3	The Education of Cyril Norwood	27
Chapter 4	The Higher Education of Boys in England	45
Chapter 5	Holding the Line?	63
Chapter 6	Marlborough and Harrow	79
Chapter 7	The English Tradition of Education	101
Chapter 8	The New World of Education	117
Chapter 9	The Norwood Report and Secondary Education	137
Chapter 10	Conclusions: The Ideal of Secondary Education	155
Notes		159
Bibliography		185
Index		193

Preface

Among the educational issues affecting policymakers, public officials, and citizens in modern, democratic, and industrial societies, none has been more contentious than the role of secondary schooling. In establishing the Secondary Education in a Changing World series with Palgrave Macmillan, the intent is to provide a venue for scholars in different national settings to explore critical and controversial issues surrounding secondary education. The series will be a place for the airing and hopefully resolution of these controversial issues.

More than a century has elapsed since Emile Durkheim argued the importance of studying secondary education as a unity, rather than in relation to the wide range of subjects and the division of pedagogical labor of which it was composed. Only thus, he insisted, would it be possible to have the ends and aims of secondary education constantly in view. The failure to do so accounted for a great deal of the difficulty with which secondary education was faced. First, it meant that secondary education was "intellectually disorientated," between "a past which is dying and a future which is still undecided," and as a result "lacks the vigor and vitality which it once possessed" (Durkheim, 1938/1987, p. 8). Second, the institutions of secondary education were not understood adequately in relation to their past that was "the soil which nourished them and gave them their present meaning, and apart from which they cannot be examined without a great deal of impoverishment and distortion" (p. 10). And third, it was difficult for secondary school teachers, who were responsible for putting policy reforms into practice, to understand the nature of the problems and issues that prompted them.

In the early years of the twenty-first century, Durkheim's strictures still have resonance. The intellectual disorientation of secondary education is more evident than ever as it is caught up in successive waves of policy changes. The connections between the present and the past have become increasingly hard to trace and untangle. Moreover, the distance between policymakers, on the one hand, and practitioners, on the other, has rarely seemed as immense as it is today. The key mission of the current series of books is, in the spirit of Durkheim, to address these underlying dilemmas of secondary education and to play a part in resolving them.

Gary McCulloch's volume explores the development of secondary education in England in the later nineteenth and early twentieth century up to World War II. This was a time of ferment for secondary education, in England as in the United States and in other parts of the world, as the provision of secondary education became established and defined itself in a drawn-out process of contestation between rival ideals and interests. Within this broader context, the educational career of Cyril Norwood stands out as he made a key contribution as a teacher, educational administrator, policymaker, and opinion maker over a period of forty years. He is both a representative and an idiosyncratic figure, and his life is a remarkable and fascinating story in its own right.

This book constitutes the first extended study of Norwood himself, a highly controversial figure at the time and since, and of the educational, social, and political ideals that he represented. It also provides an original and significant contribution to our understanding of the social relationships of secondary education as they have played themselves out over time, including the emotions of social class. The range of middle-class attitudes revolving around secondary education is analyzed in depth, from the highest academic and social aspirations to the snobberies, anxieties, and fears that were no less apparent. The rivalries between the elite independent or "public" schools and the new secondary grammar schools supported by the State are also on display in McCulloch's volume, as are the everyday experiences of pupils, schools, and reformers. The problems involved in reform are examined closely to explain both the aims and the ultimate failure of the well-known Norwood Report of 1943 on examinations and the curriculum in secondary schools. Amid all this, we can follow Norwood's own project—deeply flawed, yet passionately argued—to unite the different agencies of secondary education under the auspices of the State, drawing on the ideals of Plato, Thomas Arnold of Rugby, and Matthew Arnold. We can also relate the English case to the currents of educational and social ideas as they developed worldwide, for example, around hierarchy, equality, and education for citizenship, and the defense and mediation of established traditions and ideals, in the face of encroaching class interests and the rising global threats of fascism and war.

Cyril Norwood and the Ideal of Secondary Education is a powerful example of the kind of historical, social, and comparative accounts of secondary education that can help to explain the character and problems of secondary education in our changing world. It continues the process of helping to frame further explorations of the important issues that will characterize future volumes in this series.

Barry Franklin
Series Editor

Acknowledgments

I have incurred many debts in the research and writing for this book. Sincere thanks are due first of all to Sarah Canning and the late Ela Canning, who preserved the papers of their grandfather, Sir Cyril Norwood, with great care. They donated the papers to the University of Sheffield, where they were catalogued by the then curator of special collections and library archives, Lawrence Aspden, and included in the National Register of Archives.

I must also take this opportunity to express my grateful thanks to the Leverhulme Trust for its support for the research project "The Life and Educational Career of Sir Cyril Norwood (1875–1956)" (F/118/AU) on which this book is based, and to Dr. Colin McCaig who was the research associate on this project.

The archivists Anne Bradley (Bristol Grammar School), Geoffrey Brown (Merchant Taylors School), Rita Gibbs (Harrow School), F.H.G. Percy (Whitgift School), and T.E. Rogers (Marlborough College) have also been most helpful to me in my research. Also, many other libraries and archives have supported this work, and I would like to thank them all.

I had very useful discussions on the background of the schools that are examined in this research with Eustace Button (Bristol Grammar School), Dr. T.D.F. Money (Marlborough College), and Roger Pulbrook (Harrow School).

I had the opportunity to discuss aspects of this research through presentations at the annual conferences of the British Educational Research Association (2000), the History of Education Society (1999, 2005), and the International Standing Conference in the History of Education (2002). Many thanks to all those who took part in and commented on this work. I have also presented aspects of this work in invited staff-student seminars including most recently those held at the University of Cambridge, the University of Exeter, Manchester Metropolitan University, and the Open University.

I would like also to acknowledge that some of my previous published work cited in the bibliography has rehearsed some of the framing arguments in this study, and that the following journal articles have developed details of particular areas discussed in this book: "Cyril Norwood and the English Tradition of Education," *Oxford Review of Education*, 32/1 (2006), pp. 55–69; and "Education and the Middle Classes: The Case of the English Grammar Schools," *History of Education*, 35/6 (2006), pp. 689–704.

Barry Franklin has been most generous in setting aside his time to comment on draft chapters of the work and to generally offer advice and guidance along the way. Richard Aldrich, David Crook, and Roy Lowe have also provided very helpful suggestions.

I am very grateful to the publishers, especially Amanda Johnson, for inviting me to write this book and especially for their support and patience during the writing stage.

Most of all, may I thank my wife Sarah and my son Edward for all of their support along the way, and this book, as always, is for them.

Delegation to Canada on the "Duchess of Bedford," 1930. Norwood third from right in the front row.

Source: Norwood Family Collection

Norwood and his eldest daughter, Enid.
Source: Norwood Family Collection

Chapter 1

Introduction: Cyril Norwood and Secondary Education

Emile Durkheim, the great French sociologist and professor of pedagogy, pointed out more than a century ago the long-term dilemmas of secondary education with which it continued to wrestle.[1] Durkheim suggested that faith in the classics as the basis of secondary education had been shaken, but there was nothing new to put in its place.[2] According to Durkheim, it was history that best illustrated organizations and the ideals and aims of institutions. Moreover, he argued, the present does battle with the past, despite the fact that it derives from it and constitutes its continuation. This meant that aspects of the past were wont to disappear even when they could and should have become familiar features of the present and the future.[3]

It is appropriate to begin the present work with a reminder of Durkheim's eloquent advice because it invokes the spirit in which this is written. It addresses issues relating to secondary education as a whole through a historical analysis, on the basis of which longer-term educational issues and ideals have a significant bearing on an understanding of our present and our future.

Within this broad framework, this book has three key aims.

First, it provides an extended assessment of the life and educational career of Sir Cyril Norwood (1875–1956), one of the most prominent and influential English educators of the past century. No such appraisal has been produced before, and this omission from the literature is overdue for correction.

Second, it examines the development of secondary education in England from the 1860s through to the Education Act of 1944, which made secondary education compulsory and universal. This was a key formative phase in the historical development of secondary education, comprising a fundamental shift from a locally provided provision that was reserved to a select few, to a form of education supported by the State for a small minority of pupils, to a service for the entire age range. Yet it also embodied important underlying continuities in values and structures. It helps to explain much about the subsequent nature and problems of secondary education and indeed of society as a whole, and it is also highly instructive for the purposes of international comparison.

Third, this book tries to explain the origins, character, and appeal of an enduring ideal of secondary education that became consolidated and indeed dominant during this period, and with which Norwood himself became closely associated. It was an ideal that was expressed most fully in Norwood's books, speeches, and discussions in a range of contexts. Its frankest and most elaborate articulations, indeed, were Norwood's best known and most controversial public writings. The first of these was *The English Tradition of Education*, published in 1929, an extended defense of the English independent schools and of the potential value of the ethos that they represented in a rapidly changing educational, social, and political scene. The second was an official report produced under Norwood's chairmanship during World War II, which became known simply as the Norwood Report. This report furnished one of the most classic and forthright justifications of curriculum differentiation and school selection that has ever been published, based on the notion that there existed three types of mind that demanded different types of education for distinct roles in society. It has received widespread critical attention but deserves reassessment in the light of detailed evidence of both its longer-term and immediate context.

The Life and Career of Cyril Norwood

The central figure in this work is Cyril Norwood himself. Norwood represented both change and continuity, in his personal and family background, in his career as a teacher and a head teacher, and in his contributions to educational ideas and policies. His father, Samuel, was a Church of England priest and a grammar school head teacher who championed many of the causes with which Cyril was to become associated. Samuel's career in education failed while Cyril's was a success, but his failure, both social and educational, haunted the son throughout his life. Cyril Norwood's own educational career spanned, probably in a unique way, the whole of the period between the Acts of 1902 and 1944. In 1902, Norwood was a young grammar school classics teacher; in 1944, he was a major figure in the reconstruction of the education system. Between these dates, he was for more than four decades a leader in his field, often controversial and always vigorous in the advocacy of his ideals.

To outward appearances, Norwood's educational career was one of steady progress toward respectability and influence. He was educated from 1888 until 1894 at Merchant Taylors School in London, one of the leading Victorian public schools for the socially elite, although unusually it was principally a day school rather than a boarding school. He went on to St. John's College at the University of Oxford, where he gained outstanding first class honors in classics. He was the leading candidate for Civil Service entry and was appointed to the Admiralty. He then decided on an educational career, becoming a classics master at a long-established grammar school, Leeds Grammar School, under the new regime of the 1902 Education Act. In 1906, at the age of thirty, he was appointed headmaster of Bristol Grammar School, another school that could look back centuries to its original foundation but which was struggling to survive.

Norwood reestablished the school over the next ten years and was then appointed in 1917 as Master of Marlborough College, an elite public boarding school. In 1926, he became the head of an even more famous and prestigious school, Harrow. Finally, he returned to his old Oxford College to become its president and remained there until he retired in 1946. From 1921 until 1946, that is for a quarter of a century, he was also the chairman of the Secondary Schools Examinations Council, in which capacity he exerted a great deal of influence over the development of educational policy.

In its treatment of Norwood, at the heart of the study, this book is at least in part a biography and seeks to draw on biographical method for this purpose. It also emphasizes the relationship between the individual life and the wider society. Here it takes its cue from C. Wright Mills's *The Sociological Imagination* (1959) and its emphatic endorsement of the link to be found between biography and society. The study of individual lives has often been developed in isolation from broader considerations of historical and social dimensions. At the same time, historical and social inquiries have been liable to ignore the personal and the individual in their emphases on the bigger picture. Mills emphasized the need to try to understand the relationship between the individual and the wider structures as a key component of his ideal of the sociological imagination, which makes it possible to understand "the larger historical scene" in terms of "its meaning for the inner life and the external career of a variety of individuals."[4] Indeed, he continues, "The sociological imagination enables us to grasp history and biography and the relations between the two within society." He contends that this was a prime concern of the classic social analysts: "No social study that does not come back to the problems of biography, of history, and of their intersections within a society has completed its intellectual journey."[5] Thus, according to Mills, a key issue is to develop the capacity "to range from the most impersonal and remote transformations to the most intimate features of the human self—and to see the relations between the two."[6] Mills also traced out a key relationship, often neglected, between "personal troubles," in which an individual finds his or her values being threatened, and "public issues," involving crises in institutional arrangements.[7] The connections between biography and society thus provide a route of entry into private lives and their relationship to public careers and contributions.

These connections are exemplified also in Richard Selleck's biography of James Kay-Shuttleworth, one of the key architects of the system of elementary education developed in nineteenth-century England. Selleck argues that Kay-Shuttleworth was an "outsider" who "sought acceptance from the urban, professional middle class among whom he worked and the landed gentry into whom he married." He succeeded only partially and "became a victim of the social structure he had so sternly defended." Overall, Selleck concludes, he was "altogether more interesting than he himself, his supporters and his detractors have suggested," living "a life crowded with achievement, contradiction and distress, a fascinating mirror of the cruel age of which it was a product."[8] Similarly, Goodman and Martin, in their historical study of women educational reformers, stress the interweaving of public and private lives and suggest that "biographical research offers a particularly appropriate vehicle for exploring the lived connections between personal and political worlds."[9]

Norwood's life and career provide a further fascinating example of these issues. It is by no means a straightforward exercise to reconstruct the personal and family dimensions of Norwood's life. He was active and often prolific in communicating his ideas in public articles, books, and reports, so that his contributions to education can be traced in some detail. His personal archive, now based at the University of Sheffield, is also highly revealing in a number of respects, especially his relationship with his wife Catherine, whom he married in 1901 to begin a partnership that lasted half a century until Catherine's death in 1951. Nevertheless, in other ways Norwood's personal life is tantalizingly elusive, and it is often difficult to peer beneath his public persona that was aloof and reserved. Part of this reserve may well have been due to the circumstances of his childhood and youth. According to George Turner, Norwood's successor as Master of Marlborough College and his biographer for the *Dictionary of National Biography*, "He never spoke of his early years in a home which was darkened and impoverished by his father's intemperance: the lasting impression they made was later shown in Norwood's deep reserve, his teetotalism, special sympathy with early hardship, and the resolve that his own children should have a happy home."[10] As he grew older, this aloofness became increasingly evident.

Norwood's educational career is also interesting in that he operated across the major fault line that developed in this period between the new system of secondary education under the auspices of the State, and the established independent or "public" schools.[11] His early career was spent at day secondary schools supported by the Board of Education, Leeds Grammar School, and Bristol Grammar School. He was then the headmaster at two of the leading boarding public schools, Marlborough College and Harrow School. His position in the policy community enabled him to influence both the State and the independent sector. He did this specifically through the promotion of his ideas on the curriculum and examinations in secondary education, and more broadly through identifying the characteristics of what he defined as a distinctive "English tradition" of education.

The whole of Norwood's career was spent in educational institutions for boys or young men, staffed by males. Yet he was at the center of a major debate in the 1920s and 1930s over curriculum differentiation for boys and girls. Norwood, like many other educators of his generation, both male and female, argued that girls should receive a different curriculum from boys to suit them for a different role in society. This book will investigate the influence of this debate in the interwar period and on the longer-term development of secondary education in England.

Parity and Prestige in English Secondary Education

The period between the Education Acts of 1902 and 1944 has attracted a number of historical perspectives, although over the past generation there have been surprisingly few attempts to address this key formative phase of development within the compass of a single work. The earliest substantial study of secondary education in this period was that of John Graves, published in 1943. It set out to hold the balance between detractors and defenders of the Board of Education and its policy. According to Graves, by explaining the "peculiar difficulties" in the way of the board, it would be possible to come to a "true estimate" of both its achievements and its shortcomings.[12]

Unfortunately, no such estimate was developed, and the work generally confined itself to a description of the main features of legislation and policy developments over this time.

In many ways the most impressive single study of secondary education in this period was *Parity and Prestige in Secondary Education*, produced half a century ago by a notable sociologist of education, Olive Banks. This was concerned especially with what Banks called "the influence of the grammar school idea on the various forms of secondary education since 1902." She emphasized "those events, and in particular those controversies, which shed the most light on the social function of the various forms of secondary education, and on the sociological implications of the development of the secondary grammar school."[13] She examined the tensions between the different forms of secondary education over this period, and the persistence of the academic tradition.

Parity and Prestige was particularly focused on secondary education not in terms of the preparation that it gave for leisure or for citizenship, but as a training for certain occupational groups. As Banks concluded, "An effort has been made to place the secondary school in its relationship to the social and occupational structure of contemporary England, and to examine it, not as the purveyor of a certain type of education but as the avenue to a certain level on the social and occupational scale."[14] That is, in the interplay between social and educational factors, Banks identified a growing link between education and occupation as being of prime importance. It was change in the social and economic position of occupational groupings, Banks argued, that influenced the history of the grammar schools. Professional and administrative employment held greater prospect of social and economic rewards than technical or manual employment.[15] Thus, to Banks, the prestige and power of the academic tradition was based not on the influence of teachers and administrators, but on the vocational qualification of the academic curriculum.

Banks's work provides an important starting point for the present book, but it is lacking in several significant respects. First, it approached the topic by concentrating on general administrative and policy changes, and then it included very little biographical and case-study material to illustrate and deepen its study. Secondly, it paid scant attention to gender differences in secondary education. Thirdly, it did not attempt to relate the development of secondary education in England to international trends. And fourthly, it was based entirely on published sources rather than on archival and interview evidence. The current work will attempt to address these lacunae, while also retaining an awareness of the broader social relationships of secondary education—an awareness that Banks exemplified.

In the past thirty years, attention has shifted to the social class and gender differences reflected and reinforced in the secondary education of these years. Brian Simon's major study of the history of education in England included a critical examination of the social class inequalities of secondary education in the early decades of the twentieth century. According to Simon, "Since the outset of the century the central question in the field of public education has been the nature and scope of the secondary school system, the curriculum and means of access, the implications in terms of organization and finance, the respective share in shaping developments of central and local authorities."[16] He depicted an underlying conflict between, on the one hand, pressure to lengthen school life and to improve the quality of education for the majority of the age range, and, on the other, a determination to maintain the separate and limited system of elementary education that had been established in the Elementary Education Act of 1870.[17] In the 1920s and 1930s in particular, he argued, an official emphasis on economy had matched an elitist ideology "to programme secondary education only for an elite with, for the majority, merely ancillary provision to promote togetherness in the tasks of making capitalist industry and social relations work."[18] However, Simon acknowledged that another kind of study was required to explain why such a doctrine had "staying power" over this period, despite the existence of what he regarded as suitable conditions to abandon it.[19]

A more recent study, by Felicity Hunt, has emphasized the importance of gender as "a further, and even more fundamental category of social analysis" than that of class.[20] This work provides an account of educational decision making about schooling for girls between 1902 and 1944, especially with respect to how pupil gender became an issue in educational policy over this time.[21] Hunt documents the inconsistencies and ambiguities in the attitudes of the Board of Education toward education for girls, and the influence exerted by the board over the secondary school curriculum. In particular, she emphasizes the significance of curricular differentiation between boys and girls in the secondary schools, arguing that this reflected and maintained male privilege in the public domain.

The combination of class and gender inequalities found in Simon and Hunt supports a general interpretation, influenced by Marxist and feminist theory, that secondary education in this period was fundamentally elitist, based on entrenched social interests and ideologies, and enforced by a coterie of Board of Education officials who were, as the leading educational reformer R.H. Tawney observed, themselves drawn from an exclusive social and educational background.[22] This is a markedly different emphasis from that of Banks. Their interpretation puts the focus on the unwarranted power and influence of an unrepresentative elite group, quite unlike Banks's interpretation that explained the continued dominance of the grammar school curriculum in terms of the occupational labor market.

The Simon-Hunt framework of analysis also highlights the problematic role of the Education Act of 1902 in the development of secondary education. Rather than being credited with advancing the cause of secondary education through the introduction of a national system, albeit for a minority, the 1902 Act is now frequently castigated for its failings. Simon argued forcefully that the Act constituted a "wrong turning" and was indeed a "disaster" for the longer-term development of education over the following century. Thus, according to Simon, "the Act imposed an hierarchical and divided structure on the system of public education, thus vitiating any concept of an all-round education (including science and technology)," providing the basis for selection, classification, and differentiation of pupils at the secondary school level.[23] This view is developed further by Mel Vlaeminke, who insists that the 1902 Act "turned the clock back" and was a "profoundly retrogressive move."[24] In her view, the Act stifled earlier progress toward wider access and a broader curriculum in secondary education, and it "placed the privileges of secondary education more firmly in middle-class hands than ever before, relying on the shared understandings of a minority class to justify its policy and smooth its path, and on the carefully controlled admittance to those privileges of a tiny, mouldable minority of the majority classes."

Grammar schools, from this perspective, were given a "privileged and monopolistic position from which it would have been difficult to fail" because of the Act, and so these schools were able to accumulate prestige and establish "a model of excellence against which all other schools have since been measured."[25]

This kind of critique of the 1902 Education Act has been further strengthened through recent research on the educational initiatives and ideas of the late nineteenth century that held the promise of developing an alternative kind of system. Vlaeminke herself has documented the spread of the higher grade schools in the 1880s and 1890s, promoted by a number of local school boards as a means of going beyond the limited elementary schooling to which they were supposed to be confined under the Elementary Education Act of 1870. These schools were deemed illegal under the Cockerton judgment of 1901, and the school boards themselves were swept away in favor of local education authorities in the new legislation passed at the turn of the century. It is in the light of this that Vlaeminke concludes not only that the demise of the higher grade schools represents a "lost opportunity," but also that "the formation of the twentieth-century education system represented a huge waste of potential."[26] Similarly, Kevin Manton explores the socialist ideals of the 1880s and 1890s, discerning in this period a rich vein of morality and community that he contrasts with the post-1902 labor movement with its more "materialist" and "elitist" outlook. According to Manton, the 1902 Act "made the status accorded by education—and the socially recognised moral worth that status entails—the preserve of the middle-class minority in society."[27]

Thus, both Vlaeminke and Manton emphasize the ideals of the educational and social movements of the late nineteenth century. These are held to represent qualitatively superior values in relation to those that were to triumph as a result of the 1902 Act. Morality, community, a broader curriculum, and wider access are contrasted with materialism, elitism, class, and privilege. The current work will provide an opportunity to assess the extent to which this offers an accurate and fair representation of the ideals of the grammar schools, and the ways in which they were debated and contested, in the early twentieth century.

The work of Vlaeminke and Manton is also interesting for the way in which it links broad educational ideals to everyday practice. Vlaeminke achieves this in particular through a local case study of Bristol before and after the Education Act of 1902. Within this, she explores in detail the characteristics of specific institutions, especially St. George, Merrywood and Fairfield Road higher grade schools, and then the development of each of these schools under the new regime.[28] Manton examines the contributions of Frederick James Gould, who established a number of socialist Sunday schools and wrote lesson books for them, and Harry Lowerison, the founder of Ruskin School in Norfolk.[29] There are as yet relatively few such studies of secondary education in the early twentieth century. Again, the current work sets out to help address this deficiency through detailed case studies of Bristol Grammar School, before World War I, and of Marlborough College and Harrow School in the interwar years.

Overall, as we have seen, the historical literature has presented two sharply different types of interpretation of secondary education in the period 1902 to 1944. On the one hand, Banks highlighted the importance of the vocational labor market in an unequal society. On the other hand, Simon and Hunt have led the way toward a revisionist interpretation that emphasizes the role of the State in reinforcing and consolidating inequality in the provision of secondary education. While acknowledging the importance of these contributions, the present book sets out to explore in more detail another feature of secondary education in the early twentieth century—a feature that has hitherto attracted less attention. This is the development of a potent ideal of secondary education with what Simon described as "staying power." The ideal of secondary education was fiercely contested during this period at national, local, and institutional levels, and this book will trace the character and outcomes of this debate. It will also assess the nature of the appeal of secondary education during this time, that is, whether vocational imperatives or class and gender allegiances were principally involved, or whether other social aspirations were also responsible for the way in which secondary education established itself at this time.

Cyril Norwood was at the center of this debate in articulating the ideals of secondary education, in contesting its character both in policy and in practice, and in identifying its appeal. Much of this book will therefore be concerned with how he engaged with this debate, and with the successes and failures of this engagement. This in turn indicates a range of further social factors that were involved in the ideal of secondary education, including the search for goals such as respectability and status. The book will thus seek to explain the underlying priorities that motivated secondary education, and the reasons for its success in establishing itself over this time.

The Norwood Report

The ideal of secondary education that is most commonly associated with this period is that expressed by the Norwood Report of 1943, *Curriculum and Examinations in Secondary Schools*. This report was produced in the same month as the White Paper *Educational Reconstruction*, and, although it did not initially attract as much attention as the White Paper, it was to be of no less significance for the character of the Education Act of 1944 and of the regime of secondary education that emerged after the war.[30] The committee of twelve people that prepared the report was formally appointed as a committee of the Secondary Schools Examinations Council (SSEC), of which Cyril Norwood was the chairman, in October 1941. In spite of this, it presented its findings direct to the president of the Board of Education, R.A. Butler. Norwood was the dominant influence as well as the chairman of the committee, and the ideal of secondary education that it enunciated through its final report was in large measure his.

The introduction to the Norwood Report, brief though it was, made plain the aspirations that underlay it. The committee had been given terms of reference that were apparently narrow: "To consider suggested changes in the Secondary curriculum and the question of School Examinations in relation thereto." The report interpreted these very broadly to seek to examine the role of the curriculum in realizing the objectives of secondary schools, to debate how far examinations should be employed in assessing achievement and in what ways, and to explore the interactions between the curriculum and examinations. Still more profoundly, it posed what it called "the most fundamental question of all," which it defined as "What is the purpose of secondary education?"[31] This it tried to relate to its historical context and to rapid contemporary change, for as it declared,

> our experience as administrators, organisers and teachers has made us fully conscious that secondary education is in continual development; that it has a history, that because of its history it presents infinite variety, that it has expanded rapidly of recent years and has caught up new interests and new ideals and responsibilities, so that the very phrase "secondary education" carries with it new implications which it certainly did not carry a few years ago.

These new developments concerned not just curriculum and examinations, and so it sought to take into account broader changes and the "general trend of educational thought." For these reasons, it argued,

> In short, we were impelled to attempt to picture to ourselves, without going into details, the main features of secondary education as, judged by its past history and present tendencies, it might perhaps develop in the future; and in the sketching of that picture we have drawn on our own experience of the problems of secondary education as individuals daily engaged with them, and upon the experience of a large number of individuals and associations who have placed their views before us. This picture comprised an outline of "the conception of secondary education as a unified whole" that underlay its recommendations on the curriculum and examinations in secondary schools.[32]

Moreover, the introduction to this report contrived to address the general purpose of secondary education. Overall, it insisted, education should "help each individual to realise the full powers of his personality—body, mind and spirit—in and through active membership of a society." The curriculum should help to satisfy the intellectual, aesthetic, spiritual, and physical needs of pupils and look forward to their future lives "as citizens and as workers with hand and brain in a society of fellow citizens and fellow workers." Because personalities were so varied, the curriculum itself should be varied and flexible in order to be of maximum benefit to all individuals. Also, education should encourage the worthy and good characteristics of personalities rather than those that are unworthy and base. This meant an acknowledgment of ideals of truth, beauty, and goodness as ultimate values, which it proposed should imply "a religious interpretation of life which for us must mean the Christian interpretation of life." Thus, it held,

> education from its own nature must be ultimately concerned with values which are independent of time or particular environment, though realisable under changing forms in both, and therefore that no programmes of education which concern themselves only with relative ends and the immediate adaptation of the individual to existing surroundings can be acceptable.[33]

It also insisted that the "tradition of secondary education" should be recognized as a "fine tradition," and that its best qualities should be conserved, reinterpreted, and extended more widely to meet a new conception of secondary education.[34] It was these ideals and convictions that explained the detailed proposals that followed.

Much attention has been accorded by historians to the Norwood Report. Most of this attention has been hostile to its unscientific rationalization of a hierarchy of three types of mind, a hierarchy that would form the basis for the type of secondary school that pupils would attend.[35] This has been related to the debates surrounding the 1944 Act and its influence on the education system. But the report should also be understood in terms of the ideals and convictions that motivated it—especially those of Norwood himself; developed over the previous half century from the "experience as administrators, organisers and teachers," and the "experience of the problems of secondary education as individuals daily engaged with them." Thus a closer understanding of the Norwood Report needs to address the personal and professional background of those principally involved in formulating it, and how they had engaged with the debate over the ideal of secondary education during their own careers. This will help to explain the reasons why the Norwood Report was framed in this particular way, and also its inadequacies and failures to comprehend the rapid educational and social changes that were taking place around it.

This discussion will also help to shed light on comparative and international issues in the history of secondary education. In particular, the ideals on which the Norwood Report was based were very different from the democratic and egalitarian aspirations of the American high school over the same period.[36] The current work will highlight the distinctive historical experience that generated this starkly contrasting set of developments. At the same time, it will fashion an enhanced understanding of what such diverse systems held in common, especially as they came to face common threats in the form of fascism and war in the first half of the twentieth century. They came together under the banner of such movements as the "new education" and "education for citizenship" in order to respond to these threats. And yet even here there were differences, as the English interpretations of these international movements again proved to be distinctive.

The present work focuses therefore on the values and ideals that underlay the academic and liberal traditions of English secondary education in the late nineteenth and early twentieth centuries, to determine the extent of its achievements and failures. In general, it will illuminate the connections between individual biography, professional lives, and policy changes in the history of education through a major case study that will promote a deeper understanding of secondary education as a whole.

Chapter 2

Middle Class Education and the State

Secondary education in England in the second half of the nineteenth century was highly diverse and differentiated, but it broadly maintained a social position between the elite education of the great independent or public schools, and the new forms of mass elementary education provided under the auspices of the State. This was the heartland of English middle-class education, provided through a large number of locally endowed grammar schools. However, although it could boast fine traditions and ideals, local provision of grammar schools tended increasingly to suffer from a lack of organization and resources. The potential for the State to provide support for secondary education on a national basis attracted many, but others were resistant. In spite of the recommendations of the Taunton commission of the 1860s and the clarion calls of Matthew Arnold to organize secondary education under the auspices of the State, it was not until the Education Act of 1902 that this major step was achieved.

Revisiting the Middle Class

The leading educator Fred Clarke, in his influential treatise *Education and Social Change*, published in 1940, attributed much of the established character of the education system in England to the historical role of England's security as a nation. This included the physical security of England's island position as much as the economic security provided by the British Empire.[1] This, he argued, had encouraged not only the maintenance of historical traditions and an emphasis on continuity, but also an imitation of established approaches on the part of lower middle class and upper working class educational clienteles. Clarke noted that among these social classes, aspirations for improving their social status rested upon the educational facilities created by the Education Act of 1902. "Overwhelmingly," he suggested, "the driving force is the desire for *status* rather than for education as such."[2]

However, there is an additional issue to be considered here: it is the *insecurity* experienced by these social classes, and their anxieties around a possible decline in their position in society. Although they shared in the security that had been built up at a national level, for individuals and families there were continual dangers in the way of retaining a respectable and relatively comfortable place in society. These dangers ranged from loss of income or employment to a scandal in the family. Secondary education represented potential respite in the midst of the snakes as well as the ladders of social fortunes.

There has been much previous work on elite provision, and also on working class and mass schooling.[3] Less has been written on middle-class education, and what little has been written is generally stereotypical and unsympathetic. In his introduction to the papers from the 1981 History of Education Society annual conference, "Educating the Victorian Middle Class," Peter Searby noted the comparative neglect within the history of education of the Victorian middle class: "that huge sector of society laying between, on the one hand, those looking to the Public Schools and their like, and those on the other hand who relied on elementary schools for their children's education."[4] In the quarter of a century since then, there has been some interesting and important work around this topic, especially by W.E. Marsden and the late David Reeder.[5] Yet it remains relatively neglected, and this is also true with respect to the middle classes of the twentieth century.

Going forward from these approaches entails developing new interpretations of social class in the history of education, and of the nature of middle-class education in particular. There are signs that such a project may now be timely. The previous emphasis in the historical and sociological literature of thirty or forty years ago was on the working class and its struggles for freedom. In the past two decades we have gone through a period when an emphasis on social class was challenged: a challenge symbolized in the fall of the Soviet Union, setbacks to Marxist understandings of social class, growing awareness of other forms of social inequality and difference, and the emergence of postmodernism. Class analysis now seems to be regaining a receptive audience, with major implications for our understandings of our society and our history. At the same time, it has been reshaped by the social and political experiences of the last generation. The middle classes have grown in numbers and prominence over that time to become even more clearly a central factor in the development of society. The range of types of interest represented in the middle class, from the "old" to the "new" middle class, from the political idealism evidenced by some to the hard-nosed realism affected by others, also suggests a complexity and contestation that call out for deeper investigation. The new literature has moved forward from the monolithic view of the middle class that once was popular, to one that emphasizes a broad range and acknowledges that there is not just one middle class, there are many.

It is tempting to see the middle classes in terms of a dualism. This indeed was what R.H. Tawney suggested in his great work *Equality*, in the 1930s. Tawney argued that there were two sets of class relations at work, one rooted in the preindustrial era and the other in the economic developments of the nineteenth century:

> It is the combination of both—the blend of a crude plutocratic reality with the sentimental aroma of an aristocratic legend—which gives the English class system its peculiar toughness and cohesion. It is at once as businesslike as Manchester and as gentlemanly as Eton; if its hands can be as rough as those of Esau, its voice is as mellifluous as that of Jacob. It is a god with two faces and a thousand tongues, and, while each supports its fellow, they speak in different accents and appeal to different emotions.[6]

This argument, compellingly phrased by Tawney, evokes a fracture between older and newer ideals, and especially between the public and the grammar schools, and it is a telling one.

In the same paragraph, Tawney also went on to make a further profound insight: "Revolutionary logic, which is nothing if not rational, addresses its shattering syllogisms to the one, only to be answered in terms of polite evasion by the other. It appeals to obvious economic grievances, and is baffled by the complexities of a society in which the tumultuous impulses of economic self-interest are blunted and muffled by the sedate admonitions of social respectability."[7] Tawney thus recognized the need to comprehend the complexities of middle-class society in order to understand its character and influence, and also to address its strengths and its weaknesses. In doing so, we need to develop a greater awareness of not so much a simple dualism but also the wide range of differences within and across the middle classes, including differences that are geographical, political, cultural, social, and religious.

It is also necessary to acquire a stronger sense of the historical experience of the education of the middle classes. Outwardly, they seem to be all rational calculation, methodical, and with a steady accretion of power and influence. Inwardly, the emotions of class come to play most strikingly, in particular guilt, anxiety, and fear. Joanna Bourke's recent study *Fear: A Cultural History* explores the social and cultural dimensions of anxiety and fear, both private and public. As she suggests, increased state provision of welfare diluted fear of poverty in the early twentieth century, but fears about social status grew: "Rather than trembling about the effects of absolute privation, people shuddered to think about the consequences of *relative* impoverishment, such as being rehoused in a rougher area or forced to sell a prized possession."[8] In the United States, for example, there is much historical evidence of fear and anxiety among the middle classes. The "middling sorts" in the United States were prey to anxieties induced by their intermediate status. On the one hand, they lacked the privileges and patronage that the State might confer, while, on the other, they did not possess the education and manners of the upper classes. Thus, according to Burton J. Bledstein, they were victims of their insecure circumstances: "Caught in the middle, between great and small, the powerful and the anonymous, the dissolute above and the wretched below—the middling sorts tossed around on a sea of risk."[9] Also in relation to the United States, Stanley Aronowitz has likened the exodus from state schooling in the past few decades to "the panic in the wake of the stock market crash of 1929 or the stampede of an audience fleeing fire in a theater."[10] More generally, Peter Gay's history of the bourgeois experience of the nineteenth century—in his terms "a family of desires and anxieties"—evokes a prevailing mood that was "a mixture of helplessness and confidence," in which "endemic excitement was controlled by social devices and private defences."[11] Ambition and social climbing coexisted with nervousness at the possibility of failure or defeat.

In an exploration of the historical experience of middle-class education, it is important to search out for the experiences of particular individuals and institutions, as C.Wright Mills proposed.[12] It should also be possible to make use of case study and ethnography for particular institutions, as has been done effectively by, for example, Vlaeminke and Manton with regard to higher grade schools and socialist education.[13] In relation to the educational institutions frequented or colonized by the middle classes, A.H. Halsey's PhD thesis of 1954 with its case studies of south-west Hertfordshire and the rise of the Watford grammar schools still stands out as an example of this kind of work.[14] There is also, for example, Felicity Hunt's institutional research based on Bedford,[15] and a recent doctoral thesis by Barry Blades on Deacons School in Peterborough,[16] but surprisingly little else. Further research of this type should also raise interesting methodological challenges that require a greater engagement with the discussions about qualitative social research of the past twenty to thirty years.

There is a fundamental need, too, for historians of education to engage with the new historiography and sociology of the middle classes that has emerged in the past decade. The historical literature has developed ideas that have been drawn in many ways from F.M.L. Thompson's "rise of respectable society," embracing wide social differences, and yet focusing on "the core of the middle classes."[17] Ross McKibbin's study of classes and cultures in England from 1918 to 1951 suggests what he himself describes as a "complicated history of the middle classes in this period."[18] The research of Alan Kidd and David Nicholls on the making of the British middle class stresses the relationship between the public and private spheres in the formation of middle-class identities.[19] Leonore Davidoff and Catherine Hall, in their book *Family Fortunes: Men and Women of the English Middle Class, 1780–1850*, attempted "to reconstitute the world as provincial middle-class people saw it, experienced it and made sense of it; to accurately reconstruct an emerging culture," and produced "a rounded picture of middle-class men and women as they followed their daily pursuits and carried on their individual lives at a time when they were both agent and object of major historical change."[20] Geoffrey Crossick has focused in depth on the lower middle class, which expanded in the late Victorian period with "more numerous white-collar and minor professional employees on the one hand and more pressured shopkeepers and small producers on the other."[21] Richard Trainor's work, meanwhile, has documented in some detail the rise of a socially, occupationally, and geographically diverse urban middle class.[22]

Sociological research has similarly been active in rethinking the characteristics of social class, with a particular emphasis on the middle classes. Fiona Devine and Mike Savage, for example, have sought to relate class analysis to the idea of culture that has been influential in recent decades, an initiative that has been pursued further by Beverley Skeggs.[23] This work has set out to highlight the detailed nuances of middle-class identities and social practices, and also the ways in which middle-class families underpin the processes of social reproduction that ensure the retention of social advantages from generation to generation.[24] In this new literature, the late French sociologist Pierre Bourdieu has been widely cited to help explain the maintenance of class distinctions through the concepts of cultural capital and habitus.[25]

At the same time, there have been major signs of new interest in the middle class among sociologists of education. This has been clearly reflected, for example, in the work of Sally Power and colleagues on education and the middle class.[26] Stephen Ball's recent work has also focused on the nature of class strategies and the education market, to examine what he calls "the rhythm and murmur of middle-class voices; their changing cadences and concerns, their expression of dilemmas and ambivalences."[27] Interestingly, Ball particularly stresses the role of anxiety and fear in the middle class, a class that was "by definition a class-between, a class beset with contradictions and uncertainties" and marked by "a combination of dread and confidence."[28] Again, Sally Tomlinson suggests that by the 1980s, the middle classes, an expanding group in society, were anxious that their children should obtain the credentials for good jobs but were aware that by then policies were rationing good education, so they increasingly moved toward excluding the disadvantaged and troublesome from interfering in their children's education. According to Tomlinson, therefore,

> The educational needs of the middle classes, which by the 2000s incorporated the need to move away from meritocratic and egalitarian beliefs and exclude the disadvantaged and troublesome from interfering with their children's education, could be understood as a reaction to heightened insecurities, as global economic conditions, insecure employment, and a de-layering in public bureaucracies and private business affected middle class groups in ways they could not have envisaged up to the 1980s. The middle classes had largely become the anxious classes, certainly as far as their children's education was concerned.[29]

These are most instructive insights, but they do tend to imply that such social anxieties were novel developments, whereas the historical literature has shown clearly that the middle classes have been the "anxious classes" for at least the past two centuries.

Secondary Education in the Nineteenth Century

The ideals of secondary education that were publicly articulated in the late nineteenth century strongly emphasized its middle-class characteristics, but the ideals also made clear that they embraced a very wide range of provision. In the late 1860s, the Taunton commission produced a fine-grained social analysis of different kinds of secondary education. Thirty years later, its successor, the Bryce commission, was able to define a broad ideal of secondary education that again allowed for different kinds of institutions under its aegis. During this time, secondary education for girls was also developed with similar traits in terms of social class, but with subtly different aims and objectives from those provided for boys.

According to the Taunton Report, it was possible to define three distinct grades of education principally in terms of the amount of time that parents were willing to keep their children at school. Education that ended at about 14 years of age was the Third Grade, at about 16 Second Grade, and at 18 or 19 First Grade. These distinctions corresponded "roughly, but by no means exactly" to social gradations, particularly because those who could afford to pay more for their children's education tended in general to continue this for a longer period.[30] The First Grade was attached to the great public or independent schools and recognized the claims of the classics to a dominant place in the school curriculum. However, although many of the parents in this category were of very substantial means, there were also among them a large number who were motivated by insecurity and anxiety. These were "the great majority of professional men, especially the clergy, medical men, and lawyers; the poorer gentry; all in fact, who, having received a cultivated education themselves, are very anxious that their sons should not fall below them." Indeed, such parents had "nothing to look to but education to keep their sons on a high social level."[31] The Second Grade gave more emphasis to modern subjects, although many in this category retained a respect for Latin. These first two grades combined, in the estimation of the Taunton Report, met the demands of "all the wealthier part of the community, including not only the gentry and professional men, but all the larger shopkeepers, rising men of business, and the larger tenant farmers."[32] By contrast, the Third Grade, consisting of the smaller tenant farmers, the small tradesmen, and the superior artisans, were concerned to achieve a more basic and practical level of instruction for their children. Thus, the Taunton Report discerned in outline the middle classes as a broad and diffuse set of social gradations, and it also identified the social anxieties—no less than the ambitions—for the future that underlay their educational aspirations.

The same report also expressed clear views on the relative merits of boarding schools and day schools. Good boarding schools, it averred, were more efficient as instruments of teaching, since for its scholars "the school is the world, and the work of the school is the work of the world."[33] Pupils in such schools eschewed the distractions of the home and the family, were able to form a common understanding with the community to which they belonged, and were also in a better position to develop an appropriate character for their later responsibilities. It acknowledged that not all boarding schools lived up to these high ideals, and that indeed pupils who attended day schools, while living at home with their families, might be protected from immorality or cruelty. Day schools also had the advantage of being much less expensive, and the report recognized that more large day schools of the First Grade were needed for London and other large towns and cities. In general, however, it concluded that most parents who were in a position to choose an education of the First Grade preferred the boarding schools.

Toward the end of the century, too, the connections between secondary education and the middle classes were examined in some depth. For example, a conference on secondary education in England held at the University of Oxford in October 1893 was impressed with the need for different types of secondary education to cater to different needs. The vice chancellor of the University of Oxford, speaking at the conference, said that education should not be seen as one single ladder, but as a tree that "divides itself into several branches as it grows upwards," and therefore a number of types of secondary education were needed to suit a range of purposes. "In your zeal for one," he cautioned, "take care not to injure others."[34]

In 1895, a further official report was produced to recommend on the future of secondary education, this time under the leadership of Lord (James) Bryce. Again, this was meant to define the character of secondary education, and it developed its own ideas from the starting point provided by the earlier Taunton report. It went so far as to argue that a practical or technical education, far from being an inferior or separate type of instruction, might be understood as being one type of secondary education, a species of the genus: one branch of the tree. Indeed, it proposed, "Under the common head there are many species, each distinguished by the particular means and instruments employed and faculties exercised, but all agreeing in method and end, viz., the discipline of faculty by exercise." Secondary education itself was very broad in conception according to the Bryce Report, comprising nothing less than "education conducted in view of the special life that has to be lived with the express purpose of forming a person fit to live it."[35]

On this understanding, First Grade schools would have a special function for the formation of a learned (or literary) and professional (or cultured) class. As it continued, "This class comprehends the so-called learned professions, the ministry, law, medicine, teaching of all kinds, and at all stages, literature and the higher sciences, public life, the home and foreign civil service, and such like." It should be accessible to "capable and promising minds from every social class," and it would include both boarding and day schools.[36] In the scheme of secondary education outlined by the Bryce Report, moreover, Second Grade schools would have a particular orientation toward some form of commercial or industrial life, while Third Grade schools would train boys and girls for the higher handicrafts, or the commerce of the shop and the town.

Secondary education for girls also developed during this period, but it was viewed in terms of different ideals in relation to secondary education for boys. It had, indeed, divided aims based on gender differentiation. Advanced education was recognized as being appropriate for middle-class girls, but this was conditioned by the need to provide support for the home and the family, a function that was held to be the specific vocation of women. This set of issues created debates over whether the curriculum for girls should be oriented toward the home rather than to the university and professions, and whether there was an intellectual female elite that should be educated in the same way as the intellectual male elite.[37] These debates were also closely linked to issues of status, reflecting the relatively broad range of social backgrounds of pupils who attended the new secondary schools for girls.[38] Overall, as John Roach has commented, girls were trained "not to function as independent persons, but to become intelligent wives and mothers, more equal companions for their husbands and sons, better equipped to engage in social or voluntary work outside the home," and thus remained in an inferior social position in comparison to boys.[39]

In a general sense, then, secondary education as it developed in the second half of the nineteenth century was linked to notions of the middle classes to which it catered as a wide-ranging, amorphous, and differentiated social grouping. It was based on the idea of education as a tree as opposed to a ladder, both in relation to the types of institution to be established and in terms of the kinds of curriculum that would be appropriate. It also recognized the social anxieties of the middle classes, the fear that "their sons should not fall below them," as the Taunton Report had put it in the 1860s.[40]

Barbarians and Philistines

So far as this period is concerned, among the leading commentators, whose ideas were widely discussed, was Matthew Arnold, a major poet and critic as well as a schools inspector from 1851 until 1888. His arguments, together with the recommendations of the Taunton Report, established a basis for the development of the ideal of secondary education over the following eighty years.

Arnold's views on the urgent need for reform in schools and universities in England were strongly influenced by his travels around the continent of Europe. He was also acutely conscious of the established tradition of secondary schools, which he regarded as a rich heritage on which a new system could be built. Unlike elementary schools for the poor, which were a modern invention, he suggested that the secondary school had a long history. Indeed, he proposed, "through a series of changes it goes back, in every European country, to the beginnings of civilized society in that country; from the time when this society had any sort of organization, a certain sort of schools and schooling existed, and between that schooling and the schooling which the children of the richer class of society at this day receive there is an unbroken connection."[41] According to Arnold, this historical lineage was seen especially clearly in France, where secondary schools had been organized into a national system. By contrast, in England, secondary education suffered from a characteristic lack of organization: a system that was casual or entirely absent, a State that was powerless and indifferent, single institutions that were independent of all others and were thus places where abuses and confusion were widespread, leading to a failure to make the elements of secondary education "work fully together to a fit end," a system that was a waste of resources and produced poor results.[42]

In the place of this lack of system, Arnold insisted that a new unity was required from the middle classes, supported by a stronger role on the part of the State. The English middle class, he argued, was "cut in two in a way unexampled anywhere else."[43] On the one hand, the "professional class" was brought up with its own education that provided "fine and governing qualities," but lacking the idea of science. On the other, the "business class," increasingly important for the future, had a different form of education, of the second rank, "cut off from the aristocracy and the professions, and without governing qualities."[44] In order to unite these social groupings, it was necessary to create a public system of education, for which an education minister would be responsible. Dislike of authority had to give way to the need for an organized public system of education, which was "indispensable in modern countries." Arnold was emphatic on the importance of establishing a public system: "From the moment you seriously desire to have your schools efficient, the question between public and private schools is settled. Of public schools you can take guarantees, of private schools you cannot."[45] In turn, this would require an education minister to provide efficient administration and a central responsibility for the schools. Thus, Arnold urged, "Organise your secondary and your superior instruction."[46]

A similar argument was developed in another of Arnold's essays of this period, "A French Eton, or Middle-Class Education and the State." This set out to examine the need for a wider range of secondary schools for the middle class. Again he found inspiration from the Continent, especially in the "French Etons" such as the Toulouse Lyceum. He acknowledged that such schools were probably not of such high quality as the great English public schools such as Eton, but he pointed out that there were only five or six such schools in England, and that they were very expensive to attend. Rugby and Winchester charged fees of about £120 per year, and Eton and Harrow approaching £200 per year. They were unsuitable for the middle class: "But for the common wear and tear of middling life, and at rates tolerable for middling people, what do we produce? What do we produce at £30 a year? What is the character of the schools which undertake for us this humbler, but far more widely-interesting production?."[47]

In this essay, too, Arnold elaborated on the distinctive types of education that each social class should receive. He was in no doubt that these needed to be separate and different: "The education of each class in society has, or ought to have, its ideal, determined by the wants of that class, and by its destination."[48] Indeed, he averred,

> Society may be imagined so uniform that one education shall be suitable for all its members; we have not a society of that kind, nor has any European country. We have to regard the condition of classes, in dealing with education; but it is right to take into account not their immediate condition only, but their wants, their destination—above all, their evident pressing wants, their evident proximate destination. Looking at English society at the moment, the aim which the education of each should particularly endeavour to reach, is different.

For the highest class, Arnold suggested, school should be like a "little world," with the aim of providing "those good things which their birth and rearing are least likely to give them: to give them (besides mere book-learning) the notion of a sort of republican fellowship, the practice of a plain life in common, the habit of self-help." For the middle class, the grand aim of education should be to give "largeness of soul and personal dignity," while for the lower class, education should set out to provide "feeling, gentleness, humanity."[49]

Middle-class education should, therefore, according to Arnold, be organized so as to provide for the distinctive needs of middle-class society through schools that would charge a moderate fee, be respected for their quality, and safeguarded by regular inspection and a position in a national system of secondary schools regulated by the State. These would be "great, honourable, public institutions for their nurture—institutions conveying to the spirit, at the time of life, when the spirit is most penetrable, the solitary influences of greatness, honour, and nationality—influences which expand the soul, liberalise the mind, dignify the character."[50] Potentially, Arnold enthused, such institutions could give to the middle class what the public schools had given to the upper class: "the sense of belonging to a great and honourable public institution, which Eton and our three or four great public schools give to our upper class only, and to a small fragment broken off from the top of our middle class."[51] They might be called something like "Royal schools," supported mainly by school fees, but also by endowments and scholarships supplied by public grants. They would be directed by local bodies, with supervision from an "impartial central authority."[52]

Moreover, Arnold suggested, the development of such institutions within this kind of national system might release the energies of a middle class that was in intellectual ferment. It was the middle class that was the natural home of the "typical Englishman."[53] Furthermore, he insisted,

> It is the middle class which has real mental ardour, real curiosity; it is the middle class which is the great reader of that immense literature of the day which we see surging up all around us . . . , it is the middle class which calls it forth, and its evocation is at least a sign of a widespread mental movement in that class.[54]

The opportunity for greatness lay with the middle class, "strong by its numbers, its energy, its industry, strong by its freedom from frivolity, not by any law of nature prone to immobility of mind, actually at this moment agitated by a spreading ferment of mind." A new system of secondary education would liberalize this class with an "ampler culture," admit it to a "wider sphere of thought," help it live by larger ideas, dissipate its provincialism, cure its intolerance, and purge its pettinesses. Then, declaimed Arnold, "let the middle class rule, let it affirm its own spirit, when it has thus perfected itself."[55]

Arnold's most famous work, *Culture and Anarchy*, published in 1869, expanded upon and broadened these ideas. This major essay saw society as being divided into three basic social groups: the Barbarians (or aristocratic class), the Philistines (or middle class), and the Populace (or working class). At the same time, it acknowledged, all of this groups had something in common, for talent of some kind or another could always emerge from any of them, and they could all be led by a humane spirit. According to Arnold, they needed all to be infused with culture, which he defined as "the disinterested endeavour after man's perfection."[56] Such a study, he continued, "seeks to do away with classes; to make the best that has been thought and known in the world current everywhere; to make all men live in an atmosphere of sweetness and light, where they may use ideas, as it uses them itself, freely, nourished and not bound by them."[57] This, to Arnold, was the key social idea, and men of culture were accordingly "the true apostles of equality." Indeed, he added,

> The great men of culture are those who have had a passion for diffusing, for making prevail, for carrying from one end of society to the other, the best knowledge, the best ideas of their time; who have laboured to divest knowledge of all that was harsh, uncouth, difficult, abstract, professional, exclusive; to humanise it, to make it efficient outside the clique of the cultivated and the learned, yet still remaining the *best* knowledge and thought of the time, and a true source, therefore, of sweetness and light.[58]

Abelard, in the Middle Ages, and Lessing and Herder in eighteenth-century Germany were examples of such men of culture, according to Arnold, because they humanized knowledge, broadened the basis of life and intelligence, and, in his words, "because they worked powerfully to diffuse sweetness and light, to make reason and the will of God prevail."[59]

Moreover, Arnold argued, just as the basic social idea of culture transcended the social classes, so should the idea of the State as a means of organizing society. Personal liberty was often seen as the key ideal of English life and politics, he noted, but there was an underlying danger of a drift toward anarchy. Unlike the rest of the continent, he complained, England lacked a strong notion of the State, "the nation in its collective and corporate character, entrusted with stringent powers for the general advantage, and controlling individual wills in the name of an interest wider than that of individuals."[60] Yet this was necessary in order to counteract the trend toward anarchy. The authority of the State, however, should come not from any particular social class, but rather from the community as a whole, based on what Arnold described as our "best self."[61] This was what culture, or the study of perfection, would develop in all individuals: "We find no basis for a firm State-power in our ordinary selves; culture suggests one to us in our *best self*."[62]

In the writings of Matthew Arnold, therefore, secondary education for the middle class found its foremost prophet and its most persuasive social and educational ideals. These ideals challenged those of two very different kinds of educational institutions. On the one hand, they differed from the reformed public schools as they had developed since the early nineteenth century under the inspiration of Matthew Arnold's father, Dr Thomas Arnold. On the other, they came under pressure from an emerging type of advanced-instruction, higher-grade schools that arose from the State-organized system of elementary schooling.

Tom Brown or David Copperfield?

During the course of the nineteenth century, the public schools had grown in stature and authority to become the defining institutions of education in England. In many cases they had been in difficulties and faced widespread criticisms at the start of the century, but their provision of an elite form of education for a small minority of the population gained increasing support in succeeding decades. They were associated with a liberal education based on a classical curriculum. These provided a convenient pedagogical vehicle for adapting the philosophy of the ancient Greeks, especially Plato and Aristotle, to the nineteenth-century challenges of governing a rapidly changing mass society and an expanding global empire.[63] At the beginning of the century, the schools had been attacked for their low moral and religious standards, but by the 1870s they were praised as the training grounds for a more moral and more serious generation of future leaders of the nation and empire.[64]

The leading public school headmaster of the first half of the nineteenth century, one who personified their aspirations and ideals, was undoubtedly Thomas Arnold of Rugby School. Born in 1795, the son of a customs collector, Arnold was elected to a scholarship at Winchester College and then Corpus Christi, Oxford, before winning a fellowship at the age of nineteen at Oriel College Oxford. He was ordained as a deacon and began his own private tutorial establishment. When he applied to become the headmaster of Rugby School, at the age of thirty-two, one of his testimonials, from Dr. Hawkins, the Provost of Oriel, predicted that if he was appointed he would change the face of education throughout the public schools of England. Arnold was headmaster of Rugby from 1828 until his death in 1842, transforming the school's fortunes, but more significantly creating a legend that would spread and exert influence for many decades afterward.[65]

Arnold established a number of key features at Rugby that were imitated in public schools elsewhere, notably the authority of the headmaster, the importance of religion, a corporate ideal of the school as a whole, and the role of prefects in helping to maintain order. He insisted that the governors should not interfere with his running of the school, and also that public schools such as Rugby should remain free from state control. At the same time, he combined his role as headmaster with his position as a school chaplain, a development that was underlined by his ordination to the priesthood in 1828. This allowed him to emphasize the moral and religious dimensions of education, and to enforce strict and often severe discipline among the pupils. He preached a sermon every week in the school chapel. Religion was indeed at the center of his educational ideals. As McCrum points out, his major aims were to inculcate, first, religious and moral principles, second, gentlemanly conduct, and third, intellectual ability.[66] His own religious views often courted controversy, as he espoused a Broad Church liberalism that brought him into conflict with the Church of England hierarchy and the orthodox Anglican doctrine.[67] He was also a housemaster, for School House, and taught the Sixth Form pupils in history and the classical languages. The classics dominated the curriculum of Rugby as a whole, while the natural sciences were excluded.

Arnold's reputation grew enormously after his early death, inspiring followers in many other public schools, including, for example, C.J. Vaughan, headmaster of Harrow School from 1845 to 1859, and F.W. Farrar, master of Marlborough College from 1871 to 1876. His contribution was extolled and mythologized in a biography by A.P. Stanley, a former pupil of Rugby, and this was taken even further with a novel produced in 1857 by another former pupil, Thomas Hughes: *Tom Brown's Schooldays*. This work, perhaps the most famous of all school novels, traces Tom Brown's career at Rugby, evoking both the ideal of the English public school and the inspiration of Dr. Arnold. It emphasizes the role of the school as a community in forging the character of pupils. The tone is set early on as the narrator insists that the purpose of schools "is not to ram Latin and Greek into boys, but to make them good English boys, good future citizens, and by far the most important part of that work must be done, or not done, out of school hours."[68] Tom's father, Squire Brown, sums up his hopes for Tom at Rugby: "If he'll only turn out a brave, helpful, truth-telling Englishman, and a gentleman, and a Christian, that's all I want."[69] Tom's first sight of Rugby School evokes its traditions and authority in striking fashion:

> Tom's heart beat quick as he passed the great school field or close, with its noble elms, in which several games at foot-ball were going on, and tried to take in at once the long line of grey buildings, beginning with the chapel, and ending with the school-house, the residence of the headmaster, where the great flag was lazily waving from the highest round tower. And he began already to be proud of being a Rugby boy, as he passed the school-gates, with the oriel window above, and saw the boys standing there, looking as if the town belonged to them; and nodding in a familiar manner to the coachman, as if any of them would be quite equal to getting on the box, and working the team down street as well as he.[70]

The key figure in promoting this image of the school is, of course, Dr. Arnold himself, whether at the pulpit giving a sermon, or punishing pupils for breaching school rules. In his role as headmaster and chaplain he is the symbol of the religious and moral authority of the school: "The oak pulpit standing out by itself above the School seats. The tall gallant form, the kindly eye, the voice, now soft as the low notes of a flute, now clear and shining as the call of the light infantry bugle, of him who stood there Sunday after Sunday, witnessing and pleading for his Lord, the King of righteousness and love and glory, with whose spirit he was filled, and in whose power he spoke."[71] It was this image above all that generated and sustained the romanticized ideal of the reformed public school.

By the second half of the nineteenth century, then, the public schools had succeeded in imposing themselves as the dominant educational institutions in England. They were highly individual in their characteristics but shared in the luster of a common tradition. For all their independence, too, they were increasingly perceived in terms of a community, or even a system of schools with shared interests as well as common values. The Clarendon Commission of the 1860s included nine schools in its inquiry into public schools: Charterhouse, Eton, Harrow, Merchant Taylors, Rugby, St Paul's, Shrewsbury, Westminster, and Winchester. The claims of some other schools to standards that could get them recognized as public schools were duly corroborated by the Headmasters' Conference from 1869. In 1871, fifty schools were members of this new grouping, growing to 79 by 1886 and over one hundred by the turn of the century. As the late John Honey demonstrated, the interaction between such schools were promoted by the growing emphasis laid on sports and games, areas that received as much emphasis as academic competition. It was this that created "Tom Brown's universe," a community of schools with an established role in producing the gentlemen who comprised the social and political elite.[72]

The higher grade schools were a very different type of educational institution, emerging from the system of mass elementary instruction belatedly established by the State from the 1870s onward. Under the Elementary Education Act of 1870 and the legislation that followed, elected local school boards were made responsible for filling the gaps left by local provision in the area of elementary education. In some larger cities, school boards found an unsatisfied demand for more advanced provision and responded by providing additional facilities with a higher standard of education. In some areas, higher grade schools were able to compete with the lower grades of secondary schools for lower-middle-class pupils at a lower fee. They therefore represented a new form of opportunity for educationally able pupils from poor backgrounds, and for this reason recent historians have been increasingly sympathetic to their aspirations. According to W.E. Marsden, the higher grade schools were "one of the most relevant and forward-looking developments in English educational history."[73] Vlaeminke's study seeks to present them as prototype comprehensive schools, a modernizing phenomenon that threatened to challenge the ingrained traditions of the more established endowed schools.[74] Indeed, David Reeder and the late Brian Simon, in his final published essay, have gone so far as to portray them as a key component of an alternative system.[75]

The challenge for the local, endowed grammar schools was thus to reassert themselves as the key location of education for the middle classes, in a situation where they found themselves squeezed from above and below by the public schools and higher grade schools. The ideals of Matthew Arnold gave them a vision for such a potential resurgence. Meanwhile, they were caught most often in a more prosaic reality of substandard provision and crumbling buildings, very far removed from the grandeur of *Tom Brown's Schooldays*. It was the novelist Charles Dickens who so often evoked this social reality most effectively in his work *The Personal History of David Copperfield*, produced several years before Hughes's novel.

Dickens' novel gives a compelling account of the experience of the narrator, David Copperfield, at a school of this type. The school is Salem House, located in "a square brick building with wings, of a bare and unfurnished appearance."[76] Its schoolroom, "the most forlorn and desolate place I had ever seen," comprised

> a long room, with three long rows of desks, and six of forms, and bristling all round with pegs for hats and slates. Scraps of old copy-books and exercises litter the dirty floor. Some silk-worms' houses, made of the same material, are scattered over the desks. Two miserable little white mice, left behind by their owner, are running up and down in a fusty castle made of pasteboard and wire, looking in all the corners with their red eyes for anything to eat. A bird, in a cage very little bigger than himself, makes a mournful rattle now and then in hopping on his perch, two inches high, or dropping from it; but neither sings nor chirps.[77]

Moreover, the narrator recalls, "There is a strange unwholesome smell upon the room, like mildewed corduroys, sweet apples wanting air, and rotten books. There could not well be more ink splashed about it, if it had been roofless from its first construction, and the skies had rained, snowed, hailed, and blown ink through the varying seasons of the year."[78]

The teachers at Salem House are also suggestive of the realities of such schools. The teacher in charge, Mr. Creakle, is described as a "Tartar," both ignorant and cruel, and indeed that he was "the sternest and most severe of masters," and that, by reputation at least, "he laid about him, right and left, every day of his life, charging in among the boys like a trooper, and slashing away unmercifully."[79] Stories are told about his social background, and that he had previously been a small hop-dealer and had turned to teaching "after being bankrupt in hops, and making away with Mrs. Creakle's money."[80] Despite its poor state and the cruelty of the teachers, there is a pervasive social snobbery at Salem House. It is observed that one pupil, a coal merchant's son, was paid for against the school's coal bill and so was given the nickname of "Exchange or Barter"—"a name selected from the arithmetic book as expressing that arrangement."[81] The brittle nature of the social pretensions of the school, and of the respectability of its teachers and pupils, is exposed in an incident in which a pupil denounces a master, Mr. Mell, as a "beggar" because "his mother lives on charity in an alms-house."[82] Creakle dismisses the unfortunate Mell on the spot, sending him on his way in the full gaze of the pupils with the words: "you've been in a wrong position altogether, and mistook this for an elementary school."[83] After Mell has gone, Creakle thanks the pupil responsible "for asserting (though perhaps too warmly) the independence and respectability of Salem House."[84] The replacement teacher comes from a grammar school; the same pupil "approved of him highly, and told us he was a Brick."[85]

David Copperfield also reflects significant differences between schools. At Salem House, the narrator suggests, the boys were "generally, as ignorant a set as any schoolboys in existence; they were too much troubled and knocked about to learn; they could no more do that to advantage, than any one can do anything to advantage in a life of constant misfortune, torment, and worry."[86] He himself manages despite it all to "steadily pick up some crumbs of knowledge,"[87] but then after his mother dies he leaves to go to another school in Canterbury. The premises of this school immediately reveal a major difference from Salem House, being a grave building in a courtyard, with a "learned air about it."[88] Moreover, his new schoolmaster, Doctor Strong, is as far removed from Mr. Creakle as could be imagined. In general,

> Doctor Strong's was an excellent school, as different from Mr Creakle's as good is from evil. It was very gravely and decorously ordered, and on a sound system; with an appeal, in everything, to the honour and good faith of the boys, and an avowed intention to rely on their possession of those qualities unless they proved themselves unworthy of it, which worked wonders. We all felt that we had a part in the management of the place, and in sustaining its character and dignity.[89]

Doctor Strong is a classical scholar, always "looking out for Greek roots,"[90] and is "the idol of the whole school . . . , the kindest of men, with a simple faith in him that might have touched the stone hearts of the very urns upon the wall."[91]

The stereotypes that emerge from such fictionalized accounts are instructive for what they suggest, in dramatized fashion, about the everyday experiences of schools in the later nineteenth century.[92] In this case, they suggest the wide gulf that existed between the great public schools, "Tom Brown's universe," and the many local grammar schools that struggled to survive. According to Fred G. Walcott, when Matthew Arnold read *David Copperfield* for the first time in 1880, he recognized Mr. Creakle's school as "the type of our ordinary middle class schools."[93] They are also highly revealing in terms of the social pretensions of such schools, the educational and social aspirations that they appealed to and, by the same token, also the fears and anxieties that lay just beneath the surface. Works such as those of Thomas Hughes and Charles Dickens, which were popular and widely read, would themselves have exerted their own influence on a generation of readers, playing a part in shaping, no less than in reflecting, the social assumptions and ambitions of future pupils, teachers, and headteachers. English secondary education was born out of the political and social interests of Victorian England, and out of the persuasive powers of prophets such as Matthew Arnold. We must look to the fiction of the period to understand more deeply the dreams and nightmares of secondary education, from Rugby to Salem House, from Dr. Arnold to Mr. Creakle to Doctor Strong, as they began to take on concrete form.

Chapter 3

The Education of Cyril Norwood

This chapter provides a detailed examination of three different schools that could all be described as secondary schools from the period between 1865 and 1906: Whalley Grammar School in the 1860s and 1870s, Merchant Taylors School in the 1880s and 1890s, and Leeds Grammar School from 1901 until 1906. These schools were very different from each other in many ways, partly because of the changing context of this period as a whole, partly because of their geographical differences, and partly because of the range of secondary education for the middle classes in these years. Whalley Grammar School was a rural grammar school, Merchant Taylors School a public school based in the city of London, and Leeds Grammar School a city grammar school. Yet they all had a number of features in common. They were founded within fifteen years of each other in the mid–sixteenth century, endowed as educational institutions under Edward VI and Elizabeth I. They were all day schools, rather than boarding schools, and therefore depended on a local rather than a national clientele. They were also connected through the misfortunes and ambitions of a single family, the Norwoods. Samuel Norwood was the headmaster of Whalley Grammar School; his son, Cyril, attended Merchant Taylors School as a pupil and was later an assistant master at Leeds Grammar School. Their family history helps to illuminate the nature of secondary education and the ideals that surrounded it at this time.

Samuel Norwood and Whalley Grammar School

Whalley Abbey lay in repose near the bank of the Calder River in Lancashire, a splendid sight in a classically English rural setting. And yet it was a ruined splendor and a faded grandeur that survived into the nineteenth century, with its best and most famous years long behind it. Originally, in the fourteenth century, it had been the home of a group of Cistercian monks. The last Abbot of Whalley, John Paslew, was implicated in the rebellion (known as the Pilgrimage of Grace) against King Henry VIII and was found guilty of treason and executed in 1537. The Abbey itself was sequestered by the Crown and then sold, but changing family fortunes over the next three hundred years led to its fall into disrepair.

It was here, in the grounds of Whalley Abbey, that a small grammar school struggled to survive over the years. This school could also look back at a long history, but during the nineteenth century it fell into increasing financial difficulties. It was founded after the dissolution of the Abbey, endowed by King Edward VI in 1547, and was moved in 1725 to a new school house, with a master's house attached.[1] Its endowment was modest, and the school building somewhat dilapidated by the mid-nineteenth century, but its headmaster, the Revd George Preston, managed to keep it going for thirty-eight years from 1826 until he retired in 1864. Preston seems to have been well respected locally and was able to raise some money to improve the property of the school during his headmastership.[2] When he retired, a public meeting of the rate payers of Whalley elected as his successor a young cleric, the 24-year-old Revd Samuel Norwood.

Only six years later, the headmaster of Whalley Grammar School, still relatively new in his post, was in despair over the condition of the school. In desperation, he wrote to the Endowed School Commissioners in an urgent appeal for help. The endowment for the school from all sources, he explained, was about £28 per annum, of which £13. 6s. 8d with a few reductions was paid by the Crown. This was not sufficient to provide for necessary renovations: "The school room and premises are as ill-suited to the requirements of the present day as they can well be."[3] The school room itself was only eight feet four inches in height, and its floor was unsuitable. Besides this, he added, "the whole place is in a state of bad repair, although I have done as much as I possibly can to make it better." And so he came to the point in his letter where he says, "I wish to know if you have the power to make me a grant toward placing the school room and school premises in a satisfactory state, provided that I raise by subscription a stated sum." He concluded plaintively, "I beg to assure you, gentlemen, that nothing but the most unsatisfactory nature of the buildings and my own inability to do more for them than I have done would have induced me to trouble you."[4]

What was it that drove Samuel Norwood to write such a letter? The headmaster's predicament is an interesting and telling example of the problems of English grammar schools at that time. Their endowments were often insufficient for their needs, and the State was not prepared to intervene to support them. In that same year, 1870, the Elementary Education Act at last made it possible for locally elected School Boards to "fill the gaps" left by voluntary provision in elementary education, but this specifically excluded more advanced secondary education designed for more prosperous middle-class families. The Taunton Commission of the 1860s offered some hope with a thorough investigation of the strengths and failings of the endowed grammar schools, but it was not able to bring more financial support for them, nor a more coherent or organized pattern of provision.[5] So it was that the Endowed Schools Commission was obliged to point out in its response to Norwood's plea that it had no funds at its disposal from which to make grants to schools. It was able to only propose new schemes for educational endowments, and the endowment in this case was too small for the commission to be able to help.[6]

This was especially unfortunate as the Taunton commission had clearly identified the difficulties from which such schools characteristically suffered. A young assistant commissioner, James Bryce, later to lead his own enquiry into secondary education, provided a detailed survey of the endowed schools of Lancashire for the Taunton commission. He found a wide range of such schools scattered around the county, and that some of these had degenerated into becoming "village schools of the lowest grade."[7] Bryce categorized the endowed schools of the area into three classes. First, he identified the grammar schools in major cities such as Manchester and Preston that generally provided a classical as well as a commercial education. Second were those in the small towns and sometimes the country areas that were usually smaller than the schools in the cities but had a good income and were attended by the sons of shopkeepers and farmers. Last came grammar and other endowed schools in country areas, usually with small revenues and teaching little or no Latin, that were used mainly by the children of the laboring classes.[8]

Having visited a large number of these schools, Bryce was especially critical in his report of the state of their buildings. Most of the town schools he described as being "old, ugly, ill-ventilated, in every way offensive."[9] He was even more emphatic in his condemnation of the accommodation of the smaller grammar schools in the country areas and observed that "Of the numerous country schools there is hardly one which its trustees ought not to feel ashamed of."[10] There were several specific faults in the buildings of the country grammar schools listed in his report. They were, first of all, "ugly without and dingy within; ugly and dingy to a degree which not even a photograph could faithfully represent." He described the typical country grammar school in the following terms:

> Externally, they are plain oblong structures, with low, almost square, sometimes heavily mullioned windows, occasionally a small porch in the middle, and a bit of bare ground in front enclosed by the stone palisade so common in the northern counties. Their material is either plain brick, or more often the millstone grit or coal sandstone of the district, originally grey, but now turned almost black by damp and smoke and age. The interior is even more repulsive; the roof is low, and the small windows admit a feeble light. The walls are mostly whitewashed, or covered with a wash which once was white, but is now a grimy brown.[11]

In addition, according to Bryce, desks and benches were old, clumsy, and inconvenient, and there was a general air of discomfort and neglect. They were also unable to maintain a healthy level of warmth, as the fireplace was usually at one end of a long room, close to the teacher's desk, so that "the master is fried while the boys are frozen." The floor was usually made of stone, rather than wood, and might have been composed of even mud. The main school room was generally dirty and untidy, and the children tended to bring "the mud of the street into the room."[12]

Even worse than all of these faults, in Bryce's opinion, was the lack of proper ventilation that made them uncomfortable and malodorous. In one country school, Pilling Lane in the township of Preesall, the building was made of mud, with just one room (15 feet high, 13 1/2 feet long, and 6 3/4 feet wide) that had a rough floor composed of small stones embedded in clay and three small windows. At the time of Bryce's visit to this school, there were fifteen children in this room, and "the closeness and damp earthy smell were intolerable." In the winter there were 30 to 40 pupils. Bryce found it difficult to convey, though he tried hard, "the disgusting closeness and foulness of these small rooms packed full of boys." Moreover, he pointed out, this situation was especially unacceptable in the country schools: "In towns it is bad enough, but in country places, where a good deal of clay comes in upon boots and trousers, and where the standard of personal cleanliness is not high, the result is sometimes scarcely endurable."[13] This lack of space and ventilation, Bryce concluded, might be assumed to be "fatal to health and longevity if I had not found schoolmasters who had taught in such rooms for thirty years or more," but he insisted that "the foetid and stupefying atmosphere they had so long been breathing" caused severe harm to the education of the pupils involved.[14] Bryce acknowledged that it was difficult for the trustees and masters of the schools to repair buildings, and even more so to erect new and more suitable school premises. Finance for large-scale renovation was lacking, especially in the country districts, and with funds of their own the schools were expected to provide for themselves rather than seek support from elsewhere.

Thus, coming to a conclusion based on his inspections in Lancashire, Bryce was greatly concerned about the condition and prospects of rural endowed schools. Quite apart from their "foul and ill-scented" atmosphere, which clearly had a major effect on his attitude, he worried also about their general contribution to the education of their pupils, which was "distinctly worse than that of schools receiving Government aid." Attendance was irregular, staffing was inadequate, there was little systematic organization of classes and subjects, many old-fashioned methods and school books were still in use, and there was a general lack of good order. In some cases, too, teachers grossly neglected their duties but none was in a position to reprimand or dismiss them.[15] He was prepared to acknowledge that there were some advantages that accrued from their independence, in particular their freedom from the demands of inspectors and the State. Nevertheless, he concluded sternly, there were many schools "where abuses outweigh services—where corruption in former days and neglect continuing till now have perverted to evil the charitable intent of founders."[16] In such cases, the endowment attached to the school appeared not only useless, but actually retrogressive, since it not only maintained the school in existence but also prevented the establishment of new ones. Although they had been founded with the best of intentions, schools in this condition were now a barrier to progress:

> Deserted by the better class of people, neglected by their trustees, having lost their hold, such as it was, upon classics, and on the superior men who came to teach classics, finding it more and more difficult to withstand the demand which labour makes upon the time of children, these rural endowed schools have under-gone an irresistible though insensible change, and have sunk, in the great majority of cases, into elementary schools of the lowest grade.[17]

It was these schools therefore that provided the strongest case for active intervention in secondary education on the part of the State.

More generally, Bryce also had strong views on the social position of the endowed schools. On each side of them, he argued, there had developed two new systems of schools that catered for only one section or class of society. On the one hand, the great independent public schools were to all intents and purposes schools for the rich, while on the other, the elementary schools that were then being promoted under the auspices of the Privy Council were schools for the poor. According to Bryce, the grammar schools provided an alternative to class-based education because they were both accessible to the poor and capable of being useful to the rich. In the towns especially, he noted, their pupils were "chiefly filled by the sons of shopkeepers," but they also included the sons of artisans, and also those of professional men, merchants, and wealthy manufacturers. He therefore emphasized as a special merit the "free and universal accessibility" of grammar schools.[18] While in most schools there were "a few boys in corduroys," in many there were also "some whose parents could have afforded to send them to Eton." In general, however, Bryce commented that grammar schools tended to attract "the better people of the town." Moreover, he added, "Almost everywhere the grammar schoolmaster is held in respect, as one of the institutions of the town; the grammar school boys are proud of their square caps, and retain in after life an attachment to the place of their education."[19]

Nevertheless, Bryce also pointed out that the social position of the grammar schools showed signs of decline. This was especially true in country areas, where the sons of the squire and the clergyman were now seen less in the endowed schools than the sons of ploughmen. He suggested that the increasing class differences in the society as a whole, together with improved communications and transport that permitted greater pupil mobility over greater geographical distances, might have partly caused this trend, but he attributed it also to "the state of neglect into which many schools have been allowed to fall." Finally, he called for "a renewal of vigour and efficiency" that might even now help to recover "much of the old social prestige" that had been attached to the grammar schools.[20]

The position in which Whalley Grammar School found itself was in many respects highly characteristic of the difficulties of the country grammar schools of Lancashire, and in some ways its problems were more acute than most. Its endowment income had declined since the early years of the century, and its buildings were singled out in Bryce's general report on Lancashire as an example of the problems of disrepair.[21] A more detailed report on the school itself, based on a visit and inspection by Bryce, highlighted the issues involved. The school had no local trustees, and only Norwood himself to protect and advance its interests.[22] Norwood was also the sole teacher at the school. The supply of pupils from the local area was insufficient, partly because of the lack of a large center of population, but also because a government school took the children of the poor. Relatively few pupils therefore attended as day scholars. At the time of Bryce's visit there were six foundation pupils who were taught free under the terms of the endowment and were usually selected from the local government school. There were also eight other day boys, "the sons of small manufacturers, farmers, surgeons, and the better class of labourers," who paid £4. 4s. per year. Twenty-four boarders also attended, paying from £29. 8s. to £31. 10s., with extra payment for subjects such as French, drawing, and music.[23] The schoolroom was large enough for the purpose, but its roof was too low, and so it tended to be poorly ventilated. This was part of the master's house, which also had two dormitories for the boarders, one with fourteen beds, and the other with ten. There were grounds for concern about the standard of teaching as well as the condition of the school's infrastructure. Older pupils were taught Latin but not to a high level. In the lower part of the school, he noted, geography, history, English grammar and writing were poor, and most of the pupils were "inaccurate and altogether feeble" in their arithmetic.[24]

Also, although there was an arrangement for scholarships to Brasenose College in Oxford if there were no pupils suitable at Middleton Grammar School, none had been taken from Whalley for many years, and links appeared to have lapsed.

Bryce's recommendations for Whalley Grammar School were far-reaching. He argued that it would never be able to flourish as a day school in this quiet country neighborhood, but that it was ideally situated to develop as a boarding school: "If anything is done to establish large boarding schools on the hostel system and under public supervision this would be one of the fittest places to plant such a school."[25] This was an ambitious notion at a time when the State had no clear role in establishing or supporting either grammar or public schools. In any event, he added, there should be a board of local trustees responsible for the general management of the school, and an additional schoolroom.[26]

Thus the Revd Samuel Norwood was in a lonely and isolated position. He had no local trustees to support him, was the only teacher in his school, and was solely responsible for the maintenance of the boarders and the day scholars in his care. His only real assets were social respectability and tradition, each of which he cultivated assiduously. He was himself a respectable figure in the local community, with a BA degree from the University of London and a priesthood from the Bishop of Manchester, both awarded to him in 1867 soon after he took up his post in Whalley. In 1869 he became both a licensed preacher for the diocese of Manchester and a fellow of the Geological Society.[27] He was a keen local cricketer, and played for the Whalley Cricket Club.[28]

Norwood also had a sharp eye for the value of tradition, in particular for finding connections with well-established institutions such as the monarchy and the British Empire. He styled his school the Royal Grammar School, because of its foundational association with King Edward VI. Moreover, he demonstrated unbounded enthusiasm for India as part of the British Empire in a book that he published in 1876 titled *Our Indian Empire: The History of the Wonderful Rise of British Supremacy in Hindustan*. This was a heavily idealized and romanticized historical account designed to justify Britain's role in India. He himself described the history of the "acquisitions" of Britain's "possessions" in India as being "romantic," and he declared himself to be "convinced that his countrymen will read with pride and enthusiasm about those noble characters who have done so much to extend the borders of the British Empire, and those heroic exploits which will be related with satisfaction so long as any remnant of pride exists in the minds of Englishmen."[29] In this vein, he devoted two chapters of the book to the exploits of Robert Clive.

Norwood was at pains to explain the distinctive, even unique, character of English values and traditions, and the way in which these had been promoted in India to improve its culture and provide for its future prosperity. He insisted that England's role was entirely beneficial, unlike that of other powers such as Russia that was "only emerging from a state of semi-barbarism." By contrast, he boasted, "England has shown her power to bless as well as to subdue, to humanize as well as to conquer." Indeed, he continued, "We have bestowed all the privileges of our culture and civil order upon the myriads of the East—schools have been established, universities chartered, and the mischievous effects of 'caste' must speedily vanish before the impartial and merciful administration of justice." According to Norwood, too, the benefits of these English values were well appreciated by the local people: "Deep into the hearts and affections of the native population English rule has stricken itself: every Hindu and Bengalee can see that his property is more secure, and his life more, far more protected, under British power than they ever were even under the most honourable and merciful rajah or nabob."[30]

It was for this reason that English authority was accepted, to the extent that, as he put it, "The yea and the nay of an Englishman are respected from Cape Comorin to the Himalayas." This cultural superiority was also in his view reflected in the spread of Christianity, which was, he remarked, "slowly, but surely, making ground against the faith of Buddhism and Moslemism." He concluded with stirring certainty that in the future, "The sacred symbol of the Cross of Calvary shall be lifted up in every corner of that empire, and English rule shall yet extend wider and wider unrestrained and unchecked in its beneficent influence." The prospect was therefore "singularly bright and radiant." In general,

> In spite of many faults and many crimes, England's power has been used for blessed and holy purposes, and, when her history shall be ancient, the writer of the distant future will explain to generations yet unknown how nobly and bravely the British Empire in India was won, and how nobly and mercifully and honestly it exercised its sway.[31]

Thus, Norwood was a cultural imperialist and an apologist for prestigious institutions, comfortable and indeed complacent in the received wisdom of English traditions, in the trappings of which he sought to buttress his own position.

Yet there was also a darker side to the world of Samuel Norwood. His own family circumstances became increasingly troubled after he became head of Whalley Grammar School, and these added to the difficulties that he was already facing. In October 1868, when he was twenty-seven, his wife, Sarah, died suddenly from peritonitis.[32] He remarried only nine months later, to Elizabeth Sparks, the daughter of a banker.[33] In 1875, Elizabeth gave birth to a son, Cyril, who was to be their only child. However, virtually unknown to the outside world, they also brought up two children from Samuel's first marriage to Sarah: a girl, named Mary, and a boy, Henry. Cyril was to become well known in educational circles during the course of his lifetime, as we will see; Mary and Henry left little trace. Cyril never referred to them in public, and his obituaries after his death and the subsequent entry in the *Dictionary of National Biography* give no suggestion of their existence. Samuel's failure to acknowledge Mary and Henry in public, and Cyril's later collusion in this, may well be related to Victorian sensibilities about marriage and family. Samuel was a cleric and in a respectable position, and rumors about his first marriage and children who were not fully accounted for might well have threatened his standing in Victorian society and the Church, to say the least of it.[34]

Norwood's resignation, when it came, was dramatic enough and meant the closure of the school. In the summer of 1880, Norwood resigned his office and such was the state of the school buildings that no replacement could be appointed.[35] Confusion reigned over how best to proceed partly because there were no local trustees, a fact that Bryce had pointed out in his report to the Taunton Commission over a decade earlier. It was noted that Norwood had been appointed by a meeting of the inhabitants of the township. Some uncertainty was expressed about what this had entailed, but "It may be presumed that by the word inhabitants householders or ratepayers is meant." A report to the Endowed Schools department on the problem of Whalley Grammar School also made it clear that Norwood's discontent with the school buildings was chiefly responsible for his apparently sudden departure: "It will be seen that as long ago as 1870 Mr. Norwood made a complaint to the late Endowed Schools Commission on this subject and I understand it was principally for this reason that he resigned."[36] This report pointed out that the problems with the school buildings had never been resolved:

> The School buildings consist of a Masters house, and two main rooms with a class room. The school rooms are very low, being not much more than eight foot high. The flooring throughout the houses is in a very bad order: there is no modern school furniture, but only a few old decks and forms. The Masters house is more modern and contains several good rooms, but there is no space for Boarders.[37]

It seems that after fifteen years of struggling on his own, Norwood had simply had enough. Moreover, Bryce's earlier imaginative notion of developing the school as a boarding school was fated to remain unrealized, as the boarding accommodation, built on land that belonged to a private individual, was pulled down after Norwood's resignation.[38]

The dilemma posed by the closure of the school was highly significant. Some local people favored reopening the school as a grammar school for boys to attend during the day, while others proposed turning it into a higher class of elementary school for boys and girls paying a fee of not less than nine pence per week. It was clear that a school of some type was needed to cater to the growing population of the Whalley township, which was becoming a favorite residence for commercial people with business links in Blackburn and around east Lancashire. The notion of a higher-elementary school attracted support because "under it girls and boys would have equal educational advantages while in a Grammar School girls would be excluded." The local elementary school, which was mixed, had become overcrowded, and so "to establish a higher elementary school would be a great relief."[39] The school was reorganized in 1886 and reopened as a grammar school in 1890.[40] Samuel Norwood left the area, with his family facing an obscure and uncertain future.

Cyril Norwood and Merchant Taylors School

Among the leading independent or public schools of the nineteenth century was Merchant Taylors School in London. Its status was confirmed in the 1860s by its inclusion in the Clarendon Report as one of the nine foremost public schools in the country. According to this report, it was of the same general character as the rest of these schools, "in the antiquity of its foundation, the nature of the studies pursued in it, and its connexion with one of the ancient Universities."[41] It had been founded in 1561 by the Merchant Taylors' Company in the city of London, from the general funds of the company. Sir Thomas White, a member of the Court of Assistants of the company and also the founder of St. John's College Oxford, was one of the most influential figures in its early development through his endowment of 37 fellowships for former members of the school in the college. This established a close connection between Merchant Taylors School and the University of Oxford, a connection that was to endure for centuries. The first great headmaster of the school, soon after its foundation, was Richard Mulcaster, who was its head for twenty-five years. From the beginning, it had been a mainly classical school, with Latin taught to all pupils, Greek in the higher forms, and Hebrew also to senior pupils.[42] Like the other leading public schools, too, Merchant Taylors consisted wholly of male pupils and teachers.

The major difference between Merchant Taylors and the other schools in the Clarendon list was that this was primarily a day school, whereas the others except St Paul's were boarding. This meant that, while its curriculum and status were those of a leading public school, the cost of attending it was considerably less than that of the others. The charges and expenses for a boy at Harrow in the 1860s, for example, amounted to at least £144 per year, while in most other cases it was more than two hundred.[43] At Merchant Taylors, by contrast, there was an entrance fee of £3, with an annual fee of £10, which included all subjects taught at the school for all of its pupils, and an additional five shillings on being advanced to a higher form.[44] Some pupils stayed in boarding houses during the week, but these were not officially recognized by the school; others traveled to and from their own homes daily from around the city of London. These comparatively low pupil charges led the headmaster of the school, the Revd James Hessey, to argue in his evidence to the Clarendon Commission that it was providing an especially valuable service to the sons of clergy, physicians, surgeons, barristers, solicitors, "and others of limited and life incomes" who comprised its major clientele. Clergymen's sons and those of professionals in medicine and the law, as well as the sons of merchants and tradesmen, were especially well represented among the school's pupils.[45] Moreover, he noted, if the school moved into the country it would still be full, but it would not meet the needs of those parents who lived sufficiently near the town for their sons to attend it daily.[46]

At the same time, while Merchant Taylors School retained its established position and values in the second half of the nineteenth century, it was also undergoing some significant changes. In 1870, William Baker was appointed as the new headmaster of the school at the young age of twenty-nine. He was to have a major influence on the facilities and curriculum offered at the school in the next thirty years before he retired in 1900. The school moved from its previous home in Suffolk Lane to its new premises in Charterhouse Lane in 1875. Provision for chemistry was begun in the 1870s (physics joined the list the following decade) and a new science building was completed in 1891. The school itself grew, from 250 pupils to about 500, nearly all whom were day boys. Its academic achievements were also notable in this period, as its pupils acquired increasing numbers of prizes and scholarships to the universities. New institutions were also formed within the school at this time: the school magazine, the *Taylorian*, in 1878, school clubs for rugby, cricket, fives, and boating also in 1878, a school debating society in 1879, and a School Mission in 1890.[47]

These changes had further implications that shaped the character of the school. First, although a modern side was developed toward the end of the century, the chief master of Modern Subjects, Francis Storr, complained that it lacked status alongside the established dominance of Greek and Latin.[48] Second, the school became increasingly corporate in spirit, encouraged by the emergence of out-of-school activities in the form of societies, clubs, and games. The top hat and tailcoats worn by pupils at Suffolk Lane were abandoned as part of a gradual abolition of habits and traditions that appeared outmoded. Yet, if games and sports were embraced as a fundamental part of a public school education, academic success remained a central feature of school's ethos and wider reputation. It managed to include the corporate ethos inculcated at Rugby earlier in the century, and the cult of manliness that infused the public schools from the 1860s, while retaining its distinctive accessibility and its association with the universities, especially Oxford.

These achievements were first and foremost those of William Baker as the headmaster of the school. He was in the front rank of public school heads of the second half of the century, along with others such as Temple at Rugby School and Percival at Clifton College.[49] Nevertheless, in personal terms he was reserved and somewhat solitary. According to Draper's history of the school, "He was a shy man and he was autocratic, a combination of qualities notoriously hampering to any human intercourse." Though he was "masterful" and a "great figurehead," he remained "shy and withdrawn."[50] It was this style of headmastership that became familiar to the pupils who passed through Merchant Taylors in the final three decades of the nineteenth century.

Among the most successful pupils at the school during this time was the future classical scholar Gilbert Murray. Born in Australia, he lived with his mother in Kensington, London, at a time when the fees normally charged by public schools of the boarding type were beyond his reach. The relatively generous provision offered by Merchant Taylors made it a natural choice for him.[51] It was also very well suited to his burgeoning interest in the classics. He was especially gifted in Greek verse translation and was awarded the mark of 100 percent by the Greek examiner, Arthur Sidgwick, fellow of Corpus Christi at Oxford, for his knowledge of the *Agamemnon*. According to Murray's biographer, "At school Gilbert learned the pleasure which comes from the precision of the dead languages and the fascination of metre; and because he was able to read into the classics the values he had learned independently, the discipline strengthened the romantic, liberal spirit which came from his family background."[52] He went on to win a school scholarship to St. John's College, Oxford, which made him financially independent of his mother, except for board and lodging during the vacations. Murray gained a first class degree in the classics, leaving Oxford in 1888.

The special features of Merchant Taylors that had made it possible for Gilbert Murray to attend and to excel in the classics also help to explain the success of another of the school's notably successful pupils in this period: Cyril Norwood. After his failure at Whalley Grammar School in Lancashire, the Revd Samuel Norwood had taken his family southward, and they settled in Wanstead, Essex, near Leytonstone. Samuel Norwood found work here as a private tutor, and he was also able to give tuition to his young son. Cyril clearly showed considerable promise as a scholar from a very young age. In 1888, he was admitted to Merchant Taylors School where he was awarded a junior scholarship for the generous sum of £15. 15 s per year.[53] Two years later, he was elected to a senior scholarship, in classics, to which the even more considerable amount of £30 was attached.[54] By 1894, he was head monitor and addressed the annual gathering of the school on St. Barnabas Day, in Latin, on the events of the previous year.[55]

As a pupil at Merchant Taylors, Norwood was also notable for his prowess in sports, especially cricket. Like his father, he was an enthusiastic cricketer and made a mark both as a batsman and as a bowler. Merchant Taylors School was not particularly distinguished in this sport, and its ground at Charterhouse Square although small did not favor high scores.[56] His early performances suggested high promise as a batsman,[57] but in his final year, while his fielding improved he "failed to maintain last year's form with either bat or ball."[58] Nevertheless, he managed to retain his place in the school's First Eleven, and he also acted as secretary to the Cricket Society. His commitment to cricket reflected the fashionable cult of team sports in public schools of the time. He also played rugby football, although with less distinction.

Norwood's involvement in debating at the school was also significant, not least as an indication of his social and political views even at this early stage of his development. He made frequent contributions to the Debating Society, contributions that demonstrated an active interest in current political issues. In February 1892, for example, he spoke in favor of a motion that at present home rule would be disastrous to the interests of Ireland, arguing that home rule was "contrary to the centralising spirit of the age."[59] There was in this perhaps an interesting hint that he favored active State involvement to help bring about desirable social ends. Later in the same month, he opened a debate to support a motion that "Compulsory Sports are Desirable." In this, "He thought that the body as much as the mind should be compulsorily developed, and cited the instance of ancient Greece."[60] Such an argument, orthodox in the context of a Victorian public school, is also a reminder of the potency of the precepts as well as the languages of classical civilizations in late-nineteenth-century English elite education.

At the start of 1893, Norwood was elected secretary of the Merchant Taylors Debating Society, and he continued to put forward trenchant opinions. These views suggested an element of social concern. In a debate to consider the motion that the English workhouse system was mistaken in theory and harmful in practice, Norwood "questioned the advisability and humanity of the proposer's plan."[61] There were also clear traces of a conservative outlook on social questions. Soon afterward, he took part in a debate on the motion that "to grant so-called women's rights would be injudicious." This debate was particularly revealing of the social assumptions of public schoolboys. The opener of the debate asserted that the position of women was already sufficiently high, and that "their true place was in the home." On the other side, another pupil, "in a somewhat struggling speech," appeared to propose that women were equal to men "and desired to know why women should not have votes, a question which he seemed to be personally unable to answer." Norwood intervened to support the motion, "arguing that women's true development was to be sought in a different direction to that of men."[62] Finally, in January 1894, he is recorded as chairing a debate on the motion that "the abolition of trial by jury would be beneficial to the country." As the debate began to flag, Norwood himself became involved: "This he did by informing the House of the somewhat startling fact that the authorities invariably made a point of choosing juries from the criminal classes, perhaps in the hope that they would prove merciful to their fellow-offenders." Furthermore, "In support of this statement, he gave some family details which it is unnecessary to state here."[63] This is an intriguing suggestion that as a schoolboy he was prepared to offer insights into his own personal and family life, but the argument as a whole also conveys strongly conservative sentiments about justice and law.

Nevertheless, there is no doubt that for all his interests in other aspects of school life, the young Norwood was particularly notable as an outstanding academic prospect. In 1893, he won the Chief Hessey Divinity Prize on the classical side, General Greek Scholarship Prizes, a prize in General Latin Scholarship, and one in history and literature, as well as a prize for the work on Philology and Grammar that he undertook during a long vacation, and in English Verse.[64] The following year, when he left school, Norwood won the prestigious Sir Thomas White Scholarship to go on to St. John's College, Oxford, as well as the Tercentenary Scholarship and the Pitt Club Exhibition scholarship. The headmaster, William Baker, later remarked that Norwood had left school with more prizes than anyone "since the days of Gilbert Murray."[65] It was no surprise that he went on to win even greater honors at Oxford. He was named as "distinguished" in the examination for the Hertford Scholarship in 1895 and later gained a first class in Classical Moderations in 1896, followed by a first class in Literae Humaniores in 1898. He then gained first place in the competition for the Home and Indian Civil Service and entered the Secretariat of the Admiralty. A distinguished career clearly beckoned.

Thus, Merchant Taylors School was an example of a public school that was able to accept a wider range of pupils than others of its type, mainly from middle-class backgrounds but often with limited financial assets because of its position as a day school. It developed a corporate ethos with an emphasis on games but retained a strong academic tone, especially in the classics, which made it possible for a son of a clergyman in hard times, Cyril Norwood, to excel and succeed. It represented an ideal of secondary education that offered hope for a range of middle-class pupils, and also the possibility of adapting the established traditions of public school education in new directions.

The Case of Leeds Grammar School

Leeds Grammar School was founded in 1552, and celebrated its 350th anniversary in 1902, the year of the major Education Act.[66] Its declared goal was to prepare boys either for the universities or for a professional or commercial career.[67] For boys destined for university and the professions, it offered a classical curriculum; for those of "average ability" who were preparing to go into commerce, it provided a Modern School that taught German rather than Greek.[68] It was here that Cyril Norwood was appointed to a teaching post at the start of 1901, and he spent the next five years at this school.

How did Norwood come to take this position? He had chosen to abandon the prospect of a career in the Civil Service and had decided instead to go into teaching. The reasons for this are unclear, but it was undoubtedly a dramatic change of course. According to a testimonial from the Admiralty provided in 1905, Norwood would have had a high ranking career in the Civil Service if he had stayed, and he had already showed an ability to deal with men of affairs, administration, and prominent social and intellectual activities.[69] Although his motives for taking up a teaching position at Leeds Grammar School must remain a matter for speculation, it did provide an opportunity to follow his father Samuel into an educational career. He may well have been attracted by the promise of the impending Education Act to redeem his father's disappointments and failure.

Norwood's changing family circumstances also appear relevant in understanding his decision to move to Leeds. First, his father's decline from his earlier position of respectability and status had grown steadily more marked. He was able to continue working as a private tutor, but his health was erratic. He also appears to have sought solace in alcohol, creating severe strains within the family. Meanwhile, Henry and Mary, Cyril's brother and sister, were still living in the family home. Very few people knew of the Norwood family's secrets, but its social position was endangered and insecure especially as a result of Samuel's problems.

Second, from 1896 onward, when he was still a student at Oxford, Cyril became romantically attached to Catherine (Kitty) Kilner, a former pupil at Notting Hill and Bayswater High School.[70] Cyril and Catherine became engaged in 1897 and were eventually married on December 19, 1901. During the period of their courtship, Cyril became increasingly independent and confident in his judgments. In a long letter to Catherine in February 1897, Cyril confessed that he was "not a very interesting individual at present, being in the transition stage between manhood and boyhood."[71] He was also somewhat immature and uncertain in his ideas, developing a taste for writing verse and songs of dubious quality. He had personal doubts about the Christian faith at this time, declaring himself to be not agnostic, but not a full believer either. He also disliked dressing up to go to church, suggesting that "most men go because women go, and most women because other women do, and because they want to study the dresses."[72] He added for Catherine's benefit:

> I have a firm belief in what I can feel and believe to be true, and I have faith, thank God, in enough to make my life well worth living. I dare say it is not enough from the strictly Christian point of view, but that, I am afraid, is not the point of view which I can as yet conscientiously accept. Frankly, I cannot fully believe that Jesus was both God and Man; with full knowledge and deeper thought I shall, it may be, come to believe it. But I do believe firmly in God and the good purpose of God and our direct accountability to Him and that our work in the world is deeper than it appears, and will produce results which will influence us forever.[73]

In his later career he put such doubts behind him, although he did develop a reforming outlook within the Church of England.

Four years on, preparing for marriage, Cyril was more mature and settled in his convictions. Having found his vocation in life and his future partner, indeed, he was now prepared to deal decisively with his father. Such was the level to which Samuel had sunk that after some offence he was banned from attending his own son's wedding. In one of the very few surviving letters that make an open reference to Samuel's difficulties, Cyril explained to his wife-to-be,

> I think my father is very cut up about the wedding, but I think it will do him good. He was very glib about the defalcation he had committed, and didn't seem to think that he had done anything special. He probably doesn't, and regards himself as a man upon whom much trouble has come, and whose life is made barren. He never seems to recognise that it is he who by his own agency has made his own life barren. The only comfort is that this may be a great shock to him, and may give him the final chance.

It was also arranged that Samuel's long suffering wife, Elizabeth, would leave her husband after the wedding and go to live with the newlyweds, but Cyril was careful to suggest that "she must go somewhere first, and then come on," in order that Samuel could be "thrown off the track." Catherine agreed to tell her family that Cyril's father was not able to attend the wedding because he was ill[74] and was happy to agree to the proposed arrangement: "Your mother is so kind and has seen so much trouble that she is the last person I should mind."[75]

Cyril's new wife, Catherine, was also an important influence on his development. She was the daughter of a medical doctor, Walter Kilner, based in London. Like Cyril's family, the Kilners' social and economic position was somewhat insecure, and Catherine often noted in her letters that her father was worried about business and legal disputes.[76] She was a devout Anglican in her religious faith, working as a teacher in a Sunday school, a position she left early in 1898 because she could not control the male students.[77] Her social and political views appear to have been conservative in nature. On the London County Council elections that took place in March 1898, she wondered who would win—"progressives I am afraid."[78] Later in the same year, she expressed her satisfaction that Omburman and Khartoum had fallen to British forces, thus—in her view—avenging her childhood hero, General Gordon.[79] Catherine slowly discovered the private world of the Norwood family, including Cyril's sister: "I had no idea your sister was as old as thirty—I do not know her name—unless it is Sarah?"[80] As their relationship blossomed, she became increasingly protective of and ambitious for Cyril. Typically, as he began to prepare for his final examinations at Oxford early in 1898, she wrote, "I do wish darling you would not go in for Socker [sic], I think you have quite enough with Rugby which heaven knows is quite brutal enough, so dear Cyril do take care of yourself and don't crack up before the exam."[81] In their letters they were wont to address each other as Beatrice and Benedick, the playful courting couple in William Shakespeare's comedy *Much Ado about Nothing*. This set the tone for their marriage, which was happy and affectionate and in which she provided constant support for his ideals and ambitions. Catherine had three daughters with Cyril, and their marriage lasted for nearly fifty years until her death in August 1951. Cyril's professional life and contribution to education were much the stronger for it.

It was in these circumstances, as an independent man who had made up his mind about his future, that Norwood took up his position at Leeds Grammar School in 1901. Although his official duties only included teaching classics to the sixth form, he was so energetic that he contributed throughout the school, both in the classroom and beyond. Under able headmasterships, first of Rev J.H.D. Matthews and then from 1902 of Rev J.R. Wynne-Edwards, Norwood played a full part in helping Leeds Grammar to respond to the challenges and opportunities of the 1902 Act. He still took an active part in playing cricket[82] and also spent time coaching the junior house cricket teams.[83] He continued to have a strong interest in debating also, and, as it did at Merchant Taylors, this activity again served to provide clues to his strong and outspoken social and political views. In May 1902, taking the chair of a debate on the motion that "England has mainly herself to thank for her unpopularity abroad," he supported the opposition. According to the report on this debate in the school magazine,

> While admitting the truth of much that had been said for the motion, he still thought the real reason for England's unpopularity lay in the jealousy felt by foreign Powers. He would not defend every action of England, but he contended that the European nations had no right to throw stones at her. "If I steal a loaf," said he, "I am ready to pay the penalty, but I do not expect to be condemned by a gang of burglars."[84]

Even more startling comments were reported the following year in a debate on a motion that a trade union was beneficial to both employers and employees. Norwood, by then the honorary president of the Literary and Debating Society, spoke against the motion and was quoted in the school magazine as criticizing the "canny system and its evils," going so far as to denounce the British workman as "an idle rogue, who wanted to extort as much money out of his employer with as little labour as possible."[85] He hastily wrote a letter that was published in the next issue of the magazine to apologize for giving a "wrong impression" and to clarify his views as follows:

> I have a great respect for the working classes of England, and I do not think any such thing of them. I believe that they very naturally aim at doing a little less work and at being a little better off. In the methods they adopt to secure this I hold that they are often mistaken, but as far as the end itself goes I think that they are, Sir, very much like you and me.[86]

This was hardly a full retraction, nonetheless, and continued to suggest an underlying conservatism in his social and political outlook.

During these years, Norwood also extended his repertoire of contributions to the general life of the school. One means of doing this was through giving general lectures on history, especially of military events, and on science. In December 1901, he spoke to the Junior Literary Society about the recent events in the Boer War and the operations that had led to the relief of Ladysmith.[87] By October 1905, he had established himself sufficiently to merit the honor of presenting a lecture on Horatio Nelson on the anniversary of his death at the Battle of Trafalgar.[88] Earlier in 1905, he also read a paper to the Literary and Debating Society on "Some Wonders in Nature," which consisted of a selection of legends and stories drawn from the work of Oliver Goldsmith.[89] At the same time, he contributed to school songs, including the writing of Form 1A's new Form Song, and composing such a range of poetry for the school magazine that on his departure it dubbed him the school's "Official Bard."[90] There is in these developments an interesting echo of the educational work of his father Samuel, who had also emphasized military and imperial traditions and sought to formulate the traditions of his school in verse. Despite the difficulties experienced by the Norwood family and their increasing personal disagreements, the young Norwood clearly inherited and adapted much of his father's approach to education.

There is little evidence that at this stage Norwood had worked out a clear philosophy or ideal of secondary education, but a pair of articles contributed to the school magazine in 1904 under the title "Manent ea fata Nepotes" strongly suggest that he had identified potential developments to which he was hostile. These articles, written somewhat in the style of the then popular science fiction author H.G. Wells, follows the narrator's adventure in being transported one hundred years into the future, to Leeds Grammar School in the year 2004, and being told about the changes that have taken place in education since his own time. These changes all meet with his strong disapproval and in some cases with undisguised scorn. For example, he is informed that coeducation of boys and girls has become standard, and he expresses clear anxiety about this. The teachers are represented as serving the whims of the pupils, and the headmaster as being concerned only with the physical efficiency of the children. The control of education in this vision of the future has passed to outside inspectors, county councilors, and committees of parents.[91] The director of studies manages the intellectual pursuits of the teachers but does no teaching himself. Cricket and football have given way to mixed hockey in the winter, mixed tennis in the summer, mixed skipping between seasons, and mixed dancing all year round, due to the advances of women's rights.[92] Teachers are allowed no variation in their lessons. Greek has been abandoned, Latin survives only as a spoken language, and mathematics exists only to serve the needs of science that is taught in an elaborately heuristic method.[93] Norwood is clearly most unimpressed by these imagined changes as he returns to his own time, and concludes, " 'Awaits my descendants, does it?' I grunted, as I turned the handle of my door. 'No, that it doesn't. I'll emigrate next holidays, and marry a Polynesian.' "[94]

These intriguing articles reveal a great deal about Norwood's educational concerns at this early stage in his career. They reflect in exaggerated form nearly every prejudice and anxiety present in the secondary education of the early twentieth century. If they constituted a dream of the future, they were one of fear and insecurity. In part, the anxiety was directly educational in nature, for example, over increased local and State control over schools, heuristic methods in science, and the abolition of Greek from the curriculum. Yet it was at least as much social in its character, betraying a basic unease especially about the implications of the women's rights movement and also about the threat of a loss of authority on the part of teachers and heads in secondary schools. These fears were not expressed explicitly in terms of social class but denoted an underlying hostility to changes that would undermine the character of secondary education as it had become established through the Education Act of 1902.

Although these articles suggest that a negative impression of change as something to be avoided had already formed in Norwood's outlook on education, it is less clear that he had at this stage a full and coherent image of the kind of secondary education that should be sustained and developed further. Yet there was potentially a major contribution that schools such as Leeds Grammar School could make in the new educational regime of the 1902 Act. This was partly in providing advanced education for able pupils who could not aspire to attend the great boarding public schools. It was also in encouraging day schools such as Leeds Grammar to adopt the traditions and trappings of the boarding schools. Merchant Taylors had provided such an education for Norwood himself, and in the changing context of the early twentieth century, there was an opportunity for the city grammar schools to do the same.

Something of the challenge and opportunity confronting schools such as Leeds Grammar was indeed expressed by a visiting speaker, Dr. McGrath, the provost of Queen's College, Oxford, at the Leeds Grammar School Speech Day in October 1903. He pointed out that day schools such as Leeds Grammar were much less expensive than the so-called public schools, and that they also allowed pupils to benefit much more from "the influence of the home, and the kindly and healthy criticism of brothers and sisters," while public schools tended to discourage individuality. Parents also, McGrath went on to maintain, should be enlisted to encourage habits of punctuality and regularity.[95] Leeds Grammar School's characteristics endorsed this general picture of its potential role. In July 1905, it had on its roll 255 boys, of whom 254 were day scholars and only one was a boarder. Though its pupils were drawn mainly from families of professionals and of independent occupations (118), it also attracted children of merchants and bankers (65), clerks (44), retail traders (15), artisans (7), and a few farmers and elementary schoolmasters. They were drawn mainly from the city of Leeds itself (207), with the remainder coming from other locations around Yorkshire. The fees were a total of £13.10s. per year (substantially increased in fact from the £10. 10s. it charged the previous year).[96]

It is most likely that Norwood was aware, if only in general terms, of the ideal of secondary education represented by Leeds Grammar School and the parallels that it offered to his own schooling. He was apparently conscious of these matters in retrospect as he looked back in subsequent decades on his experience in Leeds. As he remarked privately in 1942, at Leeds he and his fellow teachers had been "all wholly wrapped up under a Headmaster of integrity and force in converting a Grammar School of the 19th Century into a great 20th Century city day school." He added, "I was certainly very happy then, newly married, Enid [his first daughter] just born. But my wife went every week to inspect the *Times* for educational vacancies, and lectured me for my intransigence. She was, as usual, quite right."[97]

Whether or not at the prompting of his wife, the popular and energetic Norwood was very active in applying for headship posts during these years, including at Bath College as early as 1902, and at leading schools in different centres.[98] Then, in 1906, aged only thirty, he was successful in obtaining a headship, at another city day school, Bristol Grammar School. One gets a further glimpse of his educational ideals at this time in his application for this headship. As he noted in this candidacy, his previous experience both as a pupil and as a teacher had been wholly in day schools, and Bristol Grammar would be very suitable insofar as it would enable him to draw on this background: "Merchant Taylor's where I spent my boyhood is one of the greatest of London day schools, but the Leeds Grammar School, where I have been for five years, holds a position in that city very similar to that held by the Grammar School in Sheffield." This was the case, according to Norwood, not only educationally but also socially, since Bristol Grammar "draws its boys from the same social stratum, and admits promising scholars from elementary schools." Moreover, he added, both Leeds and Bristol prepared their pupils not only for professional life but also for the universities: "A considerable number of boys enter upon business pursuits at sixteen, but beyond that the school feeds both the local University and the older Universities of Oxford and Cambridge." He also took the opportunity to extol the special value of day schools, as he emphasized his conviction that "this type of school more than any other is able to impart both knowledge and character, and to impress its tone without destroying individuality."[99] This invoked an ideal of the secondary school as being principally a middle-class school but one that was open to pupils from different social classes, with the aim of encouraging both individual academic ability and social qualities, and with successful pupils going on to universities and the professions. His father's school had failed to provide such an education; Merchant Taylors with the help of the Merchant Taylors Company and the Leeds Grammar School and with the support of the State were both examples of how it might be achieved.

The impressive ascent of Cyril Norwood to his own headmastership in 1906 coincided with the final eclipse of his father, Samuel. Abandoned by his family, he traveled and still looked for tutoring work, but was increasingly miserable. His sixty-fourth birthday, in July 1905, was "a sad and lonely one."[100] The following month, he lamented, "I care not how soon the curtain falls—it is all so hard and cruel."[101] In October, he celebrated the centenary of the death of Nelson—"O BRAVE NELSON"—and started to consider emigrating to Canada.[102] His spirits were lifted again briefly when he heard of his son's appointment at Bristol Grammar School, an achievement that moved him to record "Gratius deo" in his diary.[103] But he found life no easier when he arrived to settle in Canada, and thus soon thereafter he returned. He continued to talk with Cyril and to hope for a reconciliation with his own wife, but they refused to allow him further access to the family. "Little he [Cyril] seems to know or care of my deep feeling toward him or his," Samuel complained bitterly in one of his final diary entries before his death.[104]

These institutional case studies, and the family history that is entangled with them, suggest a number of major features in the ideal of secondary education in this key period of its development. These characteristics are as social as they were educational. In terms of aspirations, they reflect a strong emphasis on the importance of tradition underpinning both education and society, linking up with a view about England's preeminent place in the world. This might appear to be a complacent and conservative outlook based on security. They also had in common a holistic, corporate approach to the secondary school as an institution, embracing sporting and social activities as well as academic achievements, that is, a collective as well as an individualist ethos. Moreover, they speak of a spirit of ambition and emulation for making an improvement and for upward social mobility, to imitate the defining characteristics of the leading public boarding schools for middle-class pupils who paid a lower fee. Part of this spirit of ambition extended itself also to a willingness to bring on able pupils from poorer backgrounds who could appreciate and benefit from secondary education.

And yet, although there was an element of confidence in these social values, they were fragile and brittle in quality. Never far from the aspiration to rise in Victorian and Edwardian society was the fear of social decline, of a loss of respectability. These social anxieties were also ingrained in the ideal of secondary education at this time. It was a basic insecurity that was born out of the social differences that had been so evident in the educational reports of the 1860s, and that remained vivid at the turn of the twentieth century. They found expression in fears of falling standards and declining moral values. They were intensified and magnified by personal and family experiences, such as those of the Norwood family. The senior Norwood, an ambitious cleric, was ruined by a combination of unfortunate circumstances and personal failings. The junior Norwood made his way to outstanding individual achievement, but he still could not overcome the shadow of failure and loss. As with the Norwoods, in the domain of the ideal of secondary education too, the spirit of conservative complacency, the impulse for ambition and improvement, and the fear of social decline were to contend for dominance over the next forty years.

Chapter 4

The Higher Education of Boys in England

The school that appointed Cyril Norwood to be its headmaster, Bristol Grammar School, was a long-established school with a distinguished history, but it had major problems that required urgent resolution. The present chapter examines the nature of these problems and the contribution made by Norwood to the school's rapid growth and development during the period of his headmastership from 1906 to 1916. It will explore the expansion and social role of Bristol Grammar School over this time, and also the character of the school's curriculum. Finally, it will investigate the character of the ideals represented by Bristol Grammar School, in particular through a detailed analysis of the treatise published in 1909 by Norwood and Arthur H. Hope, *The Higher Education of Boys in England*.

The Renaissance of Bristol Grammar School

Bristol Grammar School was founded in 1532, endowed by a wealthy merchant, Robert Thorne, and with the Royal Seal of King Henry VIII. It had a chequered history and fell into steep decline toward the end of the eighteenth century before being revived in the 1840s. In 1860, a new scheme was approved to allow the school to receive pupils up to the age of nineteen, and to charge fees on a scale of £ 6 per annum for boys under 14 years of age, £ 8 for those between 14 and 16, and £ 10 for those over sixteen. However, the school was not allowed to take boarders, and for this reason a new boarding school, Clifton College, was established in 1862. Under John Caldicott, its headmaster from 1860 until 1883, Bristol Grammar increased its numbers, and in 1879 it moved to new a building in Tyndalls Park. Then, in the final decades of the nineteenth century, under a new headmaster, Robert Leighton, it fell once again into difficulties as its pupil numbers fell and competition grew. On the one hand, Clifton College rapidly emerged as a first class boarding school of the public school type. On the other, three higher grade schools were established in Bristol in the 1890s: St. George, Merrywood, and Fairfield Road. Squeezed from both sides, by the turn of the century the future of Bristol Grammar School looked increasingly bleak.[1]

Despite the new opportunities offered under the Education Act of 1902, the school remained in a parlous state. Pupil numbers were below 200, the school was in debt, and the headmaster could find no clear way forward. The city's higher-grade schools had now become municipal secondary schools, while the new role of the Board of Education seemed to create further complications with the potential of undermining the school further. For example, Leighton raised strong objections to the school's governors about the Board of Education's requirement under the new Secondary School Regulations of 1904 to allow more time in Form V for the teaching of English, on the grounds that it would reduce scope for special subjects that were needed for the purpose of university scholarships. According to Leighton indeed,

> A leading function of this School, as I conceive it, is to keep open for poor men's sons a possible road to Oxford and Cambridge. If the V forms were to give up 4½ hours per week to English, taking them from their special subjects, we could not keep these special subjects up to Scholarship standard, and no boys of ours could any longer hope for a University career.[2]

The Board of Education continued to insist on this requirement as a condition to allow the school to remain on its list of secondary schools, and the board of governors asked Leighton to comply with this.[3]

Matters came to a head when the Board of Education's inspectors arrived to inspect the school early in 1905. The school magazine noted the visit of "real live Government Inspectors," who "descended on us in the last week of March," and added sardonically, "Not being accustomed to this ordeal, we were somewhat embarrassed at finding that gentlemen so few in number could appear so ubiquitous, giving us the impression that we and our doings must certainly be of great importance to the State."[4] The inspection was certainly searching, and its final report was highly critical of the school's condition. It found only 183 day scholars on the roll as of January 1905, mainly from within Bristol. Of these, 78 were from professional and independent families, with a further 33 from the families of merchants and bankers, 25 were wards of retail traders, and 38 of clerks. The school was in a "desperate" financial position, with debts of £ 1560 1s 3d. It described the organization of the curriculum and the form system as being too complex and confusing. It also cast doubt on the contribution of the headmaster. It conceded that Leighton's ideas would benefit an "intelligent boy who goes steadily through this School," but argued that this tended to undermine the "less intelligent," going so far as to contend that Leighton himself "gives his time in School almost exclusively to the Classical Sixth Form."[5]

The report of the Board of Education also offered firm guidance on the future of the school, in particular on its position in relation to Clifton College and the newer secondary schools. As it noted,

> One view taken of the case appears to be that the School occupies an impossible position between Clifton College on the one hand and the Municipal Secondary Schools on the other, and that those who put social considerations first will, at a sacrifice, send their sons to Clifton, while those who are less careful of such considerations will send them to Fairfield Road.

Possible solutions to this dilemma included moving the school to "a safe distance from Clifton," or else changing its character to become similar to the new secondary schools. Yet the inspectors did not accept that such remedies were appropriate or necessary and argued forcefully that there remained an important role for a school such as Bristol Grammar:

> They are of opinion that there is still room for the Grammar School in Bristol, that there is, on the one hand, a large class of people who would be glad to get for £ 12 per annum the education which Clifton College gives for double that sum or more; that, on the other hand, there are many who would gladly sacrifice something to secure at the Grammar School the social advantages that they cannot obtain at the Municipal Secondary Schools.

Such "social advantages" weighed heavily with the inspectors, as they emphasized that the incentive of social considerations was just as powerful with "the lower middle class to whom Clifton College would be altogether out of reach," as it was with "the upper middle class who might just afford the Clifton fees at a sacrifice." This clear recognition of the realities of a differentiated middle class and its educational implications led the inspectors to conclude that Bristol Grammar should attempt to adapt to the new conditions in a distinctive way.[6]

The Board of Education's inspection report provided a powerful basis for the school's board of governors to insist on persevering with the development of its own approach to secondary education on the existing site. Leighton openly resisted this decision, arguing that removal to another district was inevitable and would enable the school to be once again "the best school in the neighbourhood," with "the best business."[7] Nevertheless, by now the governors had decided that the major cause of the school's problems was neither its site nor its curriculum, but the headmaster. They therefore accepted the recommendation that Leighton's "injudicious treatment of parents and an unconciliatory bearing on his part towards them," together with a "want of sympathy and tact in dealing with the boys," made it necessary to terminate his employment at the school.[8] After more than twenty years as headmaster, Leighton was forced to resign.[9]

Having dispensed with Leighton's services, the board of governors was now faced with the challenge of finding a new headmaster who could lead the school forward, increase pupil numbers, and restore its fortunes. This was no easy matter, and the process was extended when the first choice, Frederick Hillard of the Royal Grammar School in Worcester, withdrew from the appointment. It was eventually agreed to invite Cyril Norwood to take up the post, at a salary of £ 1000 per annum, with an additional capitation grant of £ 2 for every pupil registered at the school above a minimum of 200.[10] For both parties involved, this was a major gamble. Norwood was still young, and untried at this level. At the same time, Norwood was in no doubt of the difficulties surrounding this post. Friends based in Bristol were dismissive of Leighton, but they also warned him of the school's lack of money, its failure to define its role in relation to other educational institutions, and an underlying discontent among the teaching staff.[11] One close observer pointed out that the position of the grammar school was "extremely difficult," and also that the Board of Education might well be tempted to intervene more actively: "Its action in other Towns may easily be repeated in Bristol, and you may find before long that an attempt is being made to alter the character of the School."[12] Gratified as Norwood was with his appointment, the fate of Leighton will have reminded him of his own father's difficulties, the fragile nature of success, and the penalties of failure.

Norwood's response to this challenging situation was impressive by any standards. In the first assembly of the school in the new school year, attended by several governors, he pointed out the nature of the "great task": "to build up that School, to keep it what it was, to make it better—to make it, if they could, beyond dispute the first School of the city of Bristol." In doing so, he added in his speech at the school's Speech Day in October 1906, he emphasized that he wanted "to realise the ideal of a great English Day School."[13] Such was his immediate impact that by the time of the next Speech Day in 1907, the chairman of the board of governors, P.J. Worsley, could declare that "If things went on as they had begun, he thought they would have one of the great schools of England."[14]

The school's rapid advance under Norwood was vividly reflected in the growth of pupil numbers. The number of pupils enrolled increased from 184 in June 1906 to 277 in October 1907, and to 345 by January 1909. By the time of the Board of Education's next inspection of Bristol Grammar School in February 1910, numbers had doubled to 381 pupils, of whom 375 were day scholars and six were boarders.[15] In the following five years, in 1915, numbers climbed to 474 pupils, including 451 day scholars and 23 boarders.[16] The number of university scholarships also rose dramatically. Between 1907 and 1909, ten scholarships in classics, mathematics, and history at Oxford and Cambridge were gained by boys at the school. Between 1913 and 1915, 13 gained entry to Oxford, 6 to Cambridge, and 13 to the University of Bristol, with further successes at Durham, Liverpool, and St. Mary's Hospital in London.[17]

Board of Education inspectors were emphatic that it was Norwood himself who was responsible for this transformation in the position of the school. The inspectors' report in 1910 noted that Norwood when he had been appointed in 1906 had "brought with him the highest qualifications for the very difficult task with which he was faced." Since that time, it enthused, "what may, without exaggeration, be called the resuscitation of the School is without doubt his personal achievement."[18] In 1915, their report confined itself to a simple statement: "The Head Master continues to control the School with distinction and with marked success."[19]

Certainly, Norwood took every opportunity to assert his own policies and preferences. Within a few months of his arrival, he persuaded the Board of Governors to introduce a new arrangement based on four houses, wherein for each house the headmaster would appoint a captain who would be a school prefect. At the same time, a new school cap was to be designed based on a pattern and color scheme submitted by Norwood himself. For good measure, games were henceforth to be made compulsory.[20] In his annual report for 1907, Norwood declared the house system "an undoubted success," and also claimed that "since the change of the School cap I have heard no more complaints of the bad behaviour of Grammar School boys in the streets."[21] This related also to the issue of discipline that Norwood again took on under his own authority:

> Without entering into the causes of the disorder that used to be prevalent in parts of the School, I may say that I have found that by keeping control in my own hands, by not hesitating to act vigorously on occasion, and by abolishing all but slight impositions, I have been able to reduce the number of punishments greatly. It is my ambition to make the School as free from punishment as it can be, and it already compares favourably with the Schools that I have known.[22]

Such announcements reinforced the headmaster's personal involvement in a way that stamped his own authority on the school as a whole.

With the rapid growth in pupil numbers, Norwood also turned his attention to the facilities of the school, especially accommodation. By the end of 1910, as he pointed out to the school governors, the school was virtually full, and accommodation in some areas was seriously limited. The school was still in debt, but he argued strongly that it should be systematically extended, and that the City Council should be asked for the necessary funds that he estimated at between £ 6000 and £ 7000.[23] The subsequent request to Bristol's education committee to increase its grant to the school was personally signed by Norwood himself. He pointed out that although the school's income was likely to increase, its endowment needed to be used for a much larger number of boys. In this situation, he averred, "I believe the School capable of greater expansion, and because it is open to every class in the City I heartily and confidently think it worthy of increased support to enable it to fulfil its full functions in every direction." In order for the school to be able to give "the best of education in every branch of Secondary School work up to the highest limit of age," he insisted, it would require additional money.[24]

These developments also enabled the school to cultivate a clear social position and role in relation to the city of Bristol. Whereas previously it had appeared to be endangered by competitors of different types, its new expansion suggested a potentially broad clientele attracted to grammar school education of a traditional type. The inspection report of 1905, when pupil numbers were at their lowest ebb under the previous regime, showed that of the 183 day boys then attending the school, 78 were from professional and independent families, 38 from families of clerks and similar occupations, 33 from those of merchants and bankers, 25 were wards of retail trades, 6 of elementary schoolmasters, and 3 of artisans.[25] Five years later, from a group of pupils that had doubled in size in the intervening period, there were about 108 who were wards of professionals, 80 of wholesale traders, 70 of retail traders and contractors, 70 of clerks and commercial agents, 15 of public servants, 15 of artisans, 4 of farmers and 4 of laborers. While the balance remained in favor of professional and commercial backgrounds, the social class composition had perceptibly broadened toward the lower middle classes and the skilled working classes.[26] This was in the context of a highly differentiated urban Bristol society that required, according to Bristol Grammar's official history, "that brothers from those rare homes where one son went to Clifton College and another to the Grammar School should not speak if they met by chance in the street."[27]

This trend was facilitated by the provision of scholarships and the free places available under the Board of Education scheme. The school fee was about £ 12 per annum, and then £ 13 10s. as from 1908. However, under the new regulations to qualify for the maximum grant of £ 5 from the Board of Education, a grant that went to all the boys in the school between 12 and 18 years of age, the school was to provide 10 percent of its new places free. Scholarships and bursaries were raised to carry this out, based on the results of a competitive examination arranged by the headmaster.[28] By 1915, nearly 20 percent of pupils, or 92 out of 474, were paying no fees. Most of these were city or county scholars, with several of them being provided for by established Bristol Grammar scholarships.[29] By October 1911, he felt confident enough to declare at the old boys' annual dinner that "He liked to think of the Grammar School as giving people a chance who would not get that chance otherwise."[30] There was, he argued, "no real grievance at present" about entrance scholarships into the school.[31] Bristol Grammar benefited from a private bequest known as the Peloquin Scholarships, under which the governors allocated exhibitions or free places at the school to boys who had attended at least one year at a public elementary school or at the Queen Elizabeth's Hospital. Since the 1890s, between 15 and 25 boys had taken advantage of this scheme at any one time, in addition to the pupils in the Board of Education free places scheme.[32] Norwood noted that the Peloquin scholars were "drawn from a poorer social class" than those under the board's scheme.[33] The expansion of Bristol Grammar, together with its relatively low fees and the role played by the different scholarship schemes in operation, thus made it possible for its headmaster to claim that the school offered enhanced opportunity for able pupils from poorer social backgrounds.

At the same time, Norwood was concerned to improve the school's role in sending its ablest pupils onto university. He insisted that the "chief need" of the school, "acutely felt," was indeed for more leaving scholarships.[34] Commenting on the situation prevailing in 1911, Norwood said, "There are plenty of boys of ability enough to win high scholarships at Oxford and Cambridge, who cannot go up without further help. These are the boys whom I consider it good policy to assist on all grounds."[35] For this reason, he proposed that a bequest amounting to £ 900 should be used for leaving scholarships rather than for entrance scholarships. His aim was to be able to offer every year two leaving scholarships of £ 50 per year for four years each. This, he argued, would constitute "a very substantial help to the parents of our most deserving boys at a time when they most need help."[36] Additional scholarships were provided by Oxford colleges for "exceptional" success. From 1912 to 1913, Balliol College, Oxford, for the first time, gave to a group of selected boys from Bristol Grammar an open classical scholarship and an open history exhibition, each of £ 80 per year. An exhibition in classics and divinity was awarded by Brasenose College, Oxford. An anonymous donor gave £ 120 to Norwood to provide an extra leaving scholarship. Norwood took special pride also in announcing another award that came from his old college at Oxford, St. John's: "An Open Classical Scholarship of £ 100 was won at St John's College, Oxford, by a boy who came from a Bristol Elementary School when he was nearly 13 and who has developed into a scholar of considerable promise."[37] Overall from 1912 to 1913, there were 14 passes in Matriculation Examinations and a further 29 passes in the Higher Certificate of the Oxford and Cambridge Board, statistics that Norwood estimated meant that about 10 percent of the school were up to the matriculation standard for entering university.[38]

By 1915, Norwood's aim of providing two leaving scholarships per year had been achieved, in addition to two Bristol Scholarships of £ 100 each at St. John's College, Oxford, open for the first time to boys from Bristol Grammar. The total cost to the school of exhibitions, scholarships, and maintenance allowances rose from £ 620 between 1908 and 1909 to £ 948 between 1914 and 1915.[39] There are interesting similarities here to the approach taken by Norwood's old school, Merchant Taylors. Comparatively low fees combined with generous entrance scholarships had allowed a wider range of pupils from different backgrounds to attend the school; leaving scholarships and exhibitions had encouraged a high number to go on to the leading universities. Merchant Taylors School was a public school, albeit for day pupils. In the case of Bristol Grammar School, the intervention of the State had provided an opportunity for an established local grammar school to perform a similar role. The Board of Education had every reason to be satisfied that Bristol Grammar had become "a very worthy representative of what the Grammar School of a great city should be."[40] Nevertheless, it also drew on the ideals and experiences of a leading public school.

The Grammar School Curriculum

The curriculum of Bristol Grammar School also reflected these broader influences and developments. In particular, the Board of Education had a major role through the Secondary School Regulations of 1904 and subsequent circulars on individual subjects. At the same time, the curriculum and the examinations associated with it represented the importance attached to academic merit at the school. It retained an emphasis on classical subjects, although it included a wide range of other subjects as prescribed by the Board of Education's regulations. The headmaster of the school was also able to assert an important mediating role in interpreting the curriculum by providing advice and guidelines for the teachers at Bristol Grammar to follow. He also established a close relationship between the curriculum and the examination of pupils as they progressed. The school cultivated a distinctive ethos through societies and games in a way that was reminiscent of the public schools, while enforcing a strict moral code in relation to order and discipline.

Between 1909 and 1910, the school was reorganized as an institution that encouraged all pupils to be grounded in the different subjects, while enabling the strongest pupils to develop rapidly toward scholarship level. A new form, "Shell," was established to consist "mainly of scholarship boys, and others picked for promise." It operated as one form for a year, with it most successful members going on to Lower IV and the remainder entering Upper III or Lower III. This reorganization was also intended partly to meet the difficulty created by the entrance of a large number of elementary school pupils who had no background in French or Latin, and little mathematics. Norwood envisaged that such pupils would be placed either in Shell or in one of the three forms of the first year, going into the lowest set for the weak subjects.[41] The Board of Education's inspectors were suitably impressed, noting in their report on the school in 1910 that the organization of the curriculum was now "simple and efficient." The problems raised in their critical report of 1905 had been, in their view, fully addressed, and they were satisfied that "the average, or poorly endowed boy, is in no way sacrificed for the benefit of the boy of greater ability or special talents."[42] Form masters were also personally associated with the work of their own form.

Norwood also made specific arrangements for Bristol Grammar School in relation to examinations. He observed that, especially with the expansion of the school, it would become increasingly onerous, expensive, and time consuming to provide for external examination of all subjects. In return for this "labour and expense," he was unconvinced that anything was gained "save an outside impression of the School's work which may have the merit of independence." At the same time, he was highly skeptical of the value of examining younger pupils up to the age of 15: "My personal view is that the examination of very young boys is useless, and that the examination of boys up to the age of 15 is apt to be very misleading." He proposed therefore that the masters of the school should be responsible for setting examinations and should also examine and report on each other, in the same way as the masters in the public schools did. Outside examiners should be invited to examine the school in two main subjects selected by the governors, or in one main subject and two subsidiary subjects. For these purposes, Norwood defined the main subjects as being Latin, mathematics, science, and French, and the subsidiary subjects as Greek, German, geography, history, English Literature, and scripture. The remaining work would be done by the staff themselves, and examinations concluded and marked by the end of term. The governing body accepted these proposals.[43]

These provisions were highly significant for the views that they suggested about the respective roles of examinations, teachers, and the State, with implications that were far reaching for the development of secondary education over the next forty years. In the first place, while examinations were emphasized as the principal measure of individual pupil progress, their value was questioned in relation to pupils under 15. The assessment of pupils therefore required the active involvement of the teachers who were responsible for them, rather than the enforcement of uniform standards from outside the school. This meant that teachers and the school itself should be trusted to deal fairly and equitably in the interests of the pupils in their care. By the same token, it suggested that the State, while established as the key regulatory body for secondary education, should not interfere unduly with the internal processes of individual schools where they were working effectively. Norwood's rough division of the curriculum into the main and subsidiary subjects of secondary education also provides a useful indication of the priorities involved in the grammar school curriculum.

Latin was one of the subjects to which Norwood gave priority status in the curriculum. This was not surprising in view of his own specialist interests and background and the established strengths of Bristol Grammar itself. An emphasis on Latin also conformed to the approach recommended by the Board of Education. According to Circular 574, issued by the board in 1907, "a study of Latin is an essential part of a complete modern education." This was true, it argued, for a number of reasons. A knowledge of Latin was fundamental to a study of the development of European institutions, "for in it are contained the records of the development of law, religion, literature and thought." It was also an essential instrument for the use of the English language and a study of the Romance languages. The language of Latin, major authors of the classical period, and the history of Rome in this period were all therefore to be studied with care.[44]

This was certainly the case at Bristol Grammar School. The Board of Education inspection of the school in 1915 found that the position of Latin in the curriculum was "very strong." The headmaster and six assistants were responsible for its teaching. All pupils in the lower forms were given an opportunity to gain a grounding in the subject, with more intensive study from Form IVc onward. On the classical side of the school in Forms V and VI, most of the time was given to Latin and Greek, while on the Modern side about one half of the pupils continued with Latin up to the standard of the London Matriculation.[45] At the same time, Norwood was careful to point out that Bristol Grammar was not solely interested in Latin and Greek. In his speech at the school's Prize Day in October 1907, for example, he observed, "He had heard it said the Grammar School was a place where they only learnt Latin and Greek, but he could assure them that the curriculum was the ordinary one, and the one needed." Further to this, he added, the school was recognized by the Royal College of Surgeons, and he looked forward to the time when the boys would go straight from school to the medical school.[46]

The Board of Education was also emphatic about the importance of English as a subject in the secondary school. The value of both English language and English literature was emphasized, with the purpose of "training the mind to appreciate English literature," while at the same time "cultivating the power of using the English language in speech and writing."[47] It acknowledged that teachers of English should be able to develop their own methods in order to allow for individuality of treatment and for varied experiments, but also to take account of the social differences between schools. Literature was to be studied through the first-hand study of the works of great writers, which should include what it described as "the learning by heart of copious extracts from the English classics."[48] Composition was also stressed as a means to develop the adequate expression of a thought or argument, either in spoken or in written language.

At Bristol Grammar School, the headmaster took a keen interest in the teaching of English. He himself took part in the teaching and also circulated the other masters in the subject with detailed guidelines on the principles of the subject and the development of the syllabus. He pointed out, in line with the Board's advice, that "The aim of the school is to produce clear speaking and clear writing, not to attempt a detailed study of the structure of the language, nor to give a history of its literature: secondly, in the higher forms to try to form taste, and to impart a desire for good reading."[49] Norwood noted also that English lessons should not be "formal" or "stiff," but made "lively and interesting."[50] Pupils were to develop clear articulation in complete sentences, and in slow and clear reading aloud: "Boys both in repetition and in reading must be made to read clearly and with expression. They should generally stand up, and in any case speak out so as to fill the room. They may be made occasionally to criticise each other's reading."[51] So far as literature was concerned, he insisted that "what is read should be simple and great." The works of William Shakespeare and Walter Scott should, therefore, according to Norwood, form "the staple of the curriculum of every boys' school."[52] Again, the board's inspectors were effusive in their praise for the results at Bristol Grammar, while also making a number of detailed suggestions about the methods of teaching that seemed to sit uneasily with the commitment in the board's circular to encouraging individuality of treatment in different schools.[53]

History was another subject on which the Board of Education issued guidelines in this period. While again protesting that it did not wish "to lay down strict rules either as to the arrangement of the course or as to the methods of teaching," the board's advice to secondary schools was detailed and wide-ranging.[54] It suggested that the history syllabus should include both English and international history. For pupils between 12 and 16 years of age, it recommended a formal course covering the whole of English history from the invasion of the Romans to the present day. This should not attempt to treat all aspects of the history in equal detail, but to convey the overall chronological sequence of the main events, and to provide a detailed study of selected areas that would be of interest to the pupils. In most cases, it suggested, "it will be desirable to pass over with the very briefest notice those periods, the history of which is merely a record of bad government, as, e.g., the reign of Edward II, or those which are occupied with complicated and often squalid political intrigues."[55] It preferred an emphasis on events such as the Crusades, the civil war of the seventeenth century, the reign of Elizabeth, the wars for colonial supremacy, and the war of American independence. At the same time, it argued that the course should include discussion of the key events of European history that helped to explain historical developments in England, in order both to "remove the complete ignorance of any history outside England," which was too common, and to make the English history itself more intelligible and interesting.[56] It also strongly favored the idea of paying attention to the history of the town and locality in which the school was situated, by emphasizing major events that had taken place in the neighborhood, and local illustrations of general changes.

At Bristol Grammar School, the subject of history attracted particular interest from the headmaster, and the board's recommendations were interpreted in distinctive ways. Norwood circulated to his masters the key points in the board's circular on the teaching of history in secondary schools, but he pointed out that due to time constraint it was not possible to include "foreign history" in the course. He proposed that the history syllabus in the middle forms of the school should put its main emphasis on the Renaissance and Reformation, and "the exploration, adventure-spirit, and literature of [the] Elizabethan age."[57] These preferences reflect Norwood's own already established tastes for a heroic and romanticized approach to English history that brought out its patriotic endeavors and the progress made toward cultural improvement.

Norwood's approach to the history curriculum also revealed a willingness to innovate. He included a civics course in close association with the history syllabus and defended this even when the Board of Education's inspectors began to voice doubts. There was very little experience in secondary education of civics as an area of study, but Norwood was attracted to it as a potential means of teaching the importance of citizenship. He suggested that the study of civics could help first in going back to "first principles" and "showing what long words mean," and then to "understand technicalities, and to follow political discussion."[58] Few resources were available to support such a study, but Norwood commended for this purpose a recent pamphlet on the topic by C.H. Spence, head of the Modern side at Clifton College, and a systematic use of *Whitaker's Almanack*, the guide to national institutions.[59] Underlying this departure was the hope that pupils might be encouraged to understand their social and political responsibilities, in the context of a rapidly changing society, and a wider world that was becoming increasingly threatening in its outlook.

The board's inspectors were at best lukewarm about this new venture. In their inspection of the school in 1910, they noted politely that the series of lessons that had been provided in civics was "an interesting experiment," but observed that it "does not seem to be working very satisfactorily," due to the lack of a textbook, the length of the course, and also, perhaps, "the lack of a historical basis." They suggested that it might be more worthwhile to include some attention "incidentally" to civics and the more elementary facts of constitutional history as part of the history syllabus in the middle forms.[60] Five years later, they remained unconvinced, suggesting that the devotion of a whole year to civics, "however valuable in itself the study may be," tended to undermine the advanced work in history. Their report of 1915 recommended that the course in civics should be limited to one term and suggested that study of this area "might possibly gain by being limited more strictly to essentials."[61]

Nonetheless, it was clear that Norwood was determined to maintain an emphasis on preparation for the duties and responsibilities of citizenship. At the annual dinner for the old boys of the school in 1913, he declared that "They should look upon the School as a great school of citizenship—that first and foremost."[62] This expressed itself partly as a desire for social service, to support different groups in society and to reduce sectional antagonisms. It also reflected an aspiration to protect the nation and the empire against potential rivals and enemies in an increasingly volatile and hostile international environment. In both of these aspects, social service and patriotism, there was an element of anxiety present, lest the values and traditions of England be endangered by contemporary conflicts and dilemmas, whether personal, social, or political. Norwood demonstrated a commitment to social service in the local community, for example, as the president of the Bristol Young Men's Christian Association in 1914. He supported an effort to raise £ 40,000 for the building of a new center for the association and addressed large gatherings on a number of occasions in this capacity.[63] This was done, at least in part, with the aim of preventing the growth of immorality, as a local newspaper approvingly acknowledged, guiding the enthusiasm of the young man "at a time when unbridled and unguided energy may result in a permanently distorted and misshapen life."[64] Fear of decline underpinned Norwood's enthusiasm for the scouting movement, and also—as the nation moved toward the outbreak of World War I and its aftermath—for the righteous cause perceived to be represented by the Empire at war.

The Cadet Corps at Bristol Grammar School exemplified these attitudes. Norwood found the Cadet Corps "almost moribund" and "much neglected" when he came to the school and observed that it had "apparently been regarded in the School as a means of obtaining an Annual picnic in Camp."[65] He decided to take command of it himself, to take the corps out once a week, and to devise exercises in scouting, map-drawing, map-using, and signaling, as "a very useful supplement to the work of the School."[66] He had himself applied for a commission, become a captain in the first City of Bristol Volunteer Battalion Regiment of the Gloucestershire Regiment and gave up most of his summer holidays in 1907 to work with the Cadet Corps.[67] It soon became the Officers Training Corps, and its numbers grew rapidly to over two hundred in the next few years, during which a boy scouts organization for younger pupils was adopted.[68]

These developments were closely associated with the headmaster's increasing emphasis on the importance of patriotism and the defense of the nation and empire, if necessary through resort to arms. On Empire Day in 1907, he gave what was described as "a stirring address on the duties of members of such an empire as the British Empire."[69] At the School Prize Day in 1911, he made a direct connection between the Cadet Corps and citizenship, declaring that "it was the duty of every boy to make himself acquainted with military drill, he did not ask boys to join for fun, but to fit themselves for citizenship."[70] The following year, he developed this theme further in a speech on the theme of "patriotism" to the Bristol branch of the Victoria League. The colonies, he suggested, were simply the King's Dominions, "united by a spirit of brotherhood, and everyone of them in that brotherhood building up the British ideals of truth, fair dealing and equality." At the same, he also took the opportunity to emphasize the need to defend these institutions and the ideals that they represented: "As long as there was a chance of the Empire, and the country, and its ideals going down in ruins, so long it was the duty of every patriot, of an age to do it, to take his place in the line of defence."[71] Similar sentiments were expressed repeatedly following the outbreak of war. The School Prize Day in 1914 heard Norwood conclude his speech "with an appeal to everyone at the School to play their part in the war like true Englishmen and help to bring about the final victory to which we all look."[72] The War was depicted in stirring terms as a crusade for the defense of civilization: "We are fighting neither for trade nor territory, but for the freedom of the spirit and the entire hope of the future; we are fighting for something greater far than our own country. There was never a higher cause, there was never a clearer call."[73]

Rather less fervor was shown in support of science as a school subject, and indeed it attracted relatively little attention under Norwood as headmaster. It is true that new school laboratories were built during these years, but its ambition and scope seemed limited. Even the board's inspectors were moved to complain in their report of 1910 that science was not compulsory for pupils above the Upper Fourth form, and that those who did take it further were not able to attain scholarship standard. They ventured so far as to comment, "It is unfortunate that in a School of this kind there should be no individual boys, much less a Class, attempting anything like Scholarship work in Science." Such a class, they proposed, would help to stimulate science throughout the school, but, "As it is the absence of any work of a really high standard reacts unfavourably on the work and teaching as a whole, and the position of Science, notwithstanding the material improvement which has been effected in recent years, suffers by comparison with that of other important subjects of the Curriculum."[74] The inspectors of 1915 were appeased by the introduction of scholarship standard science in the Form VI on the Modern side, but there remained particular problems in physics where there was no one master who was definitely responsible for the physics work as a whole.[75]

It was principally the general corporate ethos of the school that excited the interest of the headmaster. This was again clear in relation to games and sports, and also in debating, but it was reflected vividly in the social and cultural environment of the school as well. In his application for the post of headmaster, Norwood had made a point of emphasizing his commitment to games and sports: "This side of school life is of particular importance in a day school—on the playing fields the boys cement their friendships and intensify the school feeling, while the masters deepen their acquaintance and sympathy with the boys."[76] He lost no time in making games compulsory and reorganizing these activities into four Houses. Norwood took part himself in playing cricket,[77] and by 1910, he had taken over the coaching of the school's cricket second eleven.[78] Over two hundred boys played cricket at the school in the summer term of 1909, with about the same number playing rugby football the following autumn.[79]

Debates on a range of topics were also encouraged. In 1909, the existing "Bristolians Club," which was responsible for photography, chess, and debating,[80] was dissolved and replaced by a Literary and Debating Society, whose president was the headmaster. This led to a number of activities open to the whole school, such as a debate on the motion that "votes for women is not to be advocated on grounds of public policy, nor is it to women's advantage." A general meeting held in April 1910, with the president in the chair, to which parents and friends of members of the Literary and Debating Society were invited, was attended by about two hundred pupils and others.[81] The headmaster attended debates only occasionally, but did present papers on a range of topics, including the controversy over the authorship of William Shakespeare's plays, and the strategy of the World War I. Norwood also pursued his already established predilection for composing poetry with the aim of inspiring the school's pupils, and he provided the words for a new school song, "Carmen Bristoliense," to be sung entirely in Latin.[82] The school library was well stocked with books, and the school as a whole liberally embellished with pictures, on Norwood's account, "where there was not a picture or a book in 1906."[83]

The approach taken to discipline at Bristol Grammar School under Norwood was a further indication of the overall spirit that was inculcated. Corporal punishment was retained for the most severe or difficult cases, but was to be used only by the headmaster himself. Norwood emphasized that masters must not on any account strike boys on the head, although he suggested that "with little boys it is often judicious to give them a good shake." Other forms of punishment included detention, normally for neglected homework, and punishment drills for "bad conduct, trifling, habitual untidiness, indiscipline of all kinds, unpunctuality, and slackness." Detention was regarded as more serious than drill, but the latter was still a "disgrace." Unsatisfactory work was to be reported to the pupil's form master. He instructed all masters to "repress any disorder, running, shouting, or loud talking in the corridors." Whistling was especially to be punished. This system promoted the role of the form and the form master, who was encouraged to "foster the form-feeling, and make a boy proud of his form."[84] In these ways, Norwood aimed to maintain strict order and discipline across the school, and to enable the pupils to identify with the institution as a whole.

Overall, the curriculum of Bristol Grammar School and the values that it sought to represent again tended to combine the principles of individual academic merit with the development of a team spirit that involved notions of citizenship and responsibility. It radiated a confidence in its own progress and broad utility, but it still betrayed an underlying fragility and anxiety that the ideals for which it stood might be devalued or undermined.

The Value of Secondary Education

Norwood's position as the headmaster of Bristol Grammar School, and his evident success in this role, gave him the confidence and authority to begin to expound his views on national policy issues and the future of secondary education. At his final Prize Day at the school, in October 1916, he took the opportunity to discuss education after the war, arguing that " 'bread and butter' subjects and specialization must be avoided up to sixteen years of age, since the object of education was to develop faculty of mind and body."[85] He was also increasingly involved in direct negotiations with officials of the Board of Education to improve the facilities of Bristol Grammar. Norwood's most significant contribution to national discussions about the nature and ideals of secondary education in this period, however, was embodied in the book *The Higher Education of Boys in England*, jointly edited by himself and one of his masters at Bristol Grammar School, Arthur H. Hope. This volume, published in 1909, was significant in itself for the assumptions and values that it reflected. It is also highly instructive for the insights that it gives into the early development of Norwood's ideal of secondary education, in comparison with the approach that he would develop later in his career.

In this volume, Norwood and Hope identified a distinctive national tradition of secondary education in England, rooted in the country's historical development, as a basis for understanding and addressing current issues relating to its further reform. They also attempted to compare the character of secondary education in England with the higher schools of France, Germany, and the United States. They credited William of Wykeham, the founder of Winchester College in the late fourteenth century, with establishing the principles on which boarding school life had been based. At the same time, they gave John Colet, the founder of St. Paul's School in the sixteenth century, the credit for providing "a most adequate and generously conceived prototype of what a public day school should be."[86] Colet, they contended, was the first to introduce Greek into English schools, and to also allow his high master to be a married man and a layman. Doubtless this allowed Norwood to justify his own position also as a married man and a layman.

According to Norwood and Hope, such educational institutions fell into decline after the seventeenth century, but these were revived during the nineteenth century. Secondary education "proper," they suggested, had lacked supervision until 1870 and funding until 1902, but it had at last made "a tentative effort towards efficiency and its correlated principle of national and comprehensive education."[87] They identified the Taunton commission of the 1860s, drawing attention to "scandalous shortcomings," but argued that this "great opportunity for a thorough reform" had been missed.[88] This view, consciously or otherwise, echoed the experience of Norwood's father and of Norwood himself as a child. They also highlighted the influence of Thomas Arnold at Rugby School for his major role in rehabilitating the public schools, identifying him with "the spirit of reform which has made the best of our public schools revive the old ideals of Wykeham, and has prevented the worst from altogether forgetting all ideals."[89] Essentially, their argument with respect to secondary education in general and public schools in particular was that their fortunes had improved because they had returned to their original ideals, drawing deeply on their established traditions in order to meet the new challenges of a changing society.

Following the Education Act of 1902, Norwood and Hope emphasized, the key to further development was for improved quality rather than quantity, in their view emulating Germany rather than the United States to secure "the leveling-up of the standard of existing schools rather than the spread of new institutions, to give a thorough education to the boys we have rather than a pretentious smattering to twice their number."[90] In a broader sense, however, they argued that England should follow France, Germany, and the United States by framing and evolving "a national system of education which reflects its own national genius."[91] It would therefore develop on its own distinctive and separate lines, while accepting the basic precepts that had generally come to the fore. These precepts included an increased role for state provision and state control, and Norwood and Hope were insistent on the importance of going further in this direction. The State, they averred, was "the only guide out of the wilderness." Only the State could address the complex problems of modern education at different levels, "acting through disinterested experts after consultation with the representatives of every grade and every interest," in order to effect "the emancipation of the schools."[92] They suggested that this further growth of state authority might be exercised in the future through the agency of the nine inspectorial districts that were then in existence, so that the Board of Education itself could devote its attention to larger issues. It would, for example, be able to outline the different curricula that were suitable for different kinds of schools and the ways in which each subject should develop. This would allow the nurturing and what Norwood and Hope described as "soundly educational curricula" that would provide for "the necessary unity of treatment for each subject throughout the course." The ultimate effect, they optimistically concluded, would be to "inspire and guide our masters," rather than to "reduce them to machines."[93] Overall, while the immediate duty was to "check abuses and remedy defects," the real task of the State was "to guide and inspire a living system of schools, each of which shall retain its own personality and develop its own talent, but without misdirection of energy or uncritical acceptance of tradition."[94]

In putting forward these recommendations, Norwood and Hope were embellishing the general arguments that Matthew Arnold had expressed in the late nineteenth century. Secondary education could only prosper and be fulfilled through the enhanced role of the State, guided by experts in educational issues. They were also careful to insist that this increase in central control should not be at the expense of the independence of individual schools, headmasters, or teachers. The board, they noted, should destroy the "isolation" but not the "independence" of schools.[95] Within the wide limits fixed by the board, "the freer the headmaster and his staff can be left the better will it be for the school."[96] Furthermore, they added, the headmaster should remain a powerful figure, both within the school, and in relation to external control in "a position of undisputed supremacy, responsibility, and prestige."[97] The headmaster should still be able to appoint all staff and also to dismiss them with cause. However, if the governing body of the school supported the appeal of a dismissed master, this would normally lead to the headmaster's resignation, "for it would be inconceivable that he would be able to conduct the school with the necessary authority and support when he would count among his staff the very man whom he had endeavoured to dismiss, and a majority who had condemned his action."[98] Nevertheless, they continued to hope that, as they put it,

> in every English Secondary School there will be one man to be found whose word is law, and whom every one within the building shall unhesitatingly follow; who shall have full liberty to build up, within and without the walls, a school which shall reflect his ideals, and be, so far as he can make it, "himself writ large"; who shall concede to his colleagues their own spheres, and their own liberty, of action; who shall inspire them, and be inspired by them, working always in sympathy and charity.[99]

If Matthew Arnold was the source of their ideal of the State, it was his father, Thomas Arnold, who inspired their vision of the role of the headmaster.

At the same time, they were also sympathetic to the cause of assistant masters, arguing that they should be properly trained, remunerated, and protected with due rights if they were to develop into a professional body. Schoolmasters, they proposed, should have a State-aided pension scheme, as was already the case in France and Germany. They should also develop a stronger union to voice their opinions and protests and to safeguard their professional interests. Under the direction of the State, Norwood and Hope enthused, school teachers could become more akin to civil servants, united on a national basis and across different kinds of school. Indeed, they urged, it was the most urgent of all reforms to improve the situation of teachers, so that they might become "experts and enthusiasts, with a proper professional status, and with a career of public status guaranteed and rewarded by the only agency that can give adequate expression to public needs, and bring system into public enterprise."[100] Once again, as in other arenas, in their view, the key to further progress lay with and through the State.

Their ideas about the future organization of secondary education were also clearly developed. They proposed that there should be two main types of secondary school, "differing not in spirit or in method or, more than is inevitable, in social status, but rather in the subjects taught and in the life-work for which they prepare."[101] The first type, roughly corresponding to a combination of the German *Gymnasium* and *Realgymnasium*, would aim at keeping its boys to the age of 18 or 19, and then sending them either through the universities into the professions, or directly into the higher technical institutes and special schools. This species of secondary school would be represented by the greater grammar schools and the public school. The second type, akin to the German *Realschule*, would give its pupils a liberal education up to the age of 16, to train them for local commerce and industry, and would be developed in the smaller grammar schools and the municipal and county schools. Such a distinction might not be fully tenable for some time, in the authors' view, but in the future "it seems probable that two main types of school, one classical and semi-classical, the other frankly modern, will prove most effective, each concentrating its energy upon its own definite work."[102]

This argument harked back to the ideas of the Taunton Commission of the 1860s about the differentiation of secondary schools, although it sought also to forge a clearer connection between the leading grammar schools and the elite public schools. Norwood and Hope saw these two types of schools tending to converge, as the grammar school was likely to become "the increasingly predominating type of English higher education, by reason of its combination of liberal study and contact with life," while the public school needed to be "brought into line and given the leading place it merits in a national system."[103] On this view, grammar schools—both large and small—would draw increasingly on the established ethos of the public schools, while the public schools would improve their intellectual standards and break down their caste spirit. They recognized also that there could be a form of preparation of a secondary level that would be appropriate for the purposes of training for commerce and industry. Nevertheless, they continued to regard secondary education as being the preserve of a minority of the age range, with the elementary schools retaining their established role as the majority. They also persisted in focusing exclusively on boys and failing to mention, let alone address, the needs of girls.

Norwood and Hope were convinced that such measures would ensure that the value of secondary education, as a fundamental requirement for all technical and superior study, would become increasingly clear to parents and employers. It was important, they argued, for individuals and for the nation as a whole, especially as it offered what they regarded as "an opportunity for bringing different classes together when they are young and generous, and thereby introducing a spirit of union and mutual sympathy into a nation so split to-day by faction that it is risking both its internal prosperity and its outward strength."[104] Not content with this, however, they concluded their work by indulging their "fantasy" of what an English higher school for boys might look like in the future:

> The ideal school will be situated on the outskirts of an English town, in the fairest suburb, where the houses are merging into the open country. If there is a hill, it will stand upon it, in order that from the class-room window its boys may look over the city, with its mills and steeples, away to the farms and moorland beyond, with perhaps a glimpse of the sea. Its buildings will be both noble and useful, with an individuality of their own, that they may be loved, like a person. There shall be a great hall, where the boys can pray together and sing together—songs of the school, of the town, of the nation. Some one may have built a lofty tower, with a deep bell, to summon to work and play, so that the Old Boys, busy down in the town, may see the tower, and hear the notes of the bell, and remember. . . . It will also be a school of Englishmen that shall make their country's name spell honour and her empire service.[105]

This was also represented as an ideal that transcended and overcame class barriers and cultural differences. On the one hand, "The son of the Prime Minister will sit side by side with the son of the merchant and the son of the workman, who has been given a scholarship because the country needs so badly all the talent she can discover." On the other, "The Indian judge's boy will be there as a boarder, perhaps because his father has been a boy there himself and learnt there to know and love all sorts and conditions of men."[106] Boys would fight for their side on the playing fields and would work keenly in the classroom.

This sentimental and roseate image drew copiously on the successful revival of Bristol Grammar School and the personal experience of the man widely regarded as its second founder, Cyril Norwood. It had a strong and broad appeal, as its expansion reflected, and yet it also rested on insecure foundations, and the extent of its expansion was limited. The war was a major test to the ideals that the Bristol Grammar School represented, and in the aftermath of the war, threats to this civilization were to become stronger than before.

Chapter 5

Holding the Line?

Following the end of World War I, and throughout the 1920s, the Board of Education was involved in a number of debates that centered on the character and ideals of secondary education. These included issues around the secondary school curriculum, including the role of the board itself, the position of teachers, and the extent of gender differentiation. At the same time, the further development of secondary education became increasingly controversial, especially in the light of a pamphlet published by the Labour Party that argued in favor of extending secondary education to the whole age range.[1] Cyril Norwood played a significant role in these policy debates, for he was by now a national figure in education. His position was consolidated even before the end of the war when he was appointed master of Marlborough College, one of the leading public boarding schools in the country. He was also invited to be a member of the new Secondary Schools Examinations Council (SSEC), which reported to the Board of Education. Norwood's influence, while it supported reform in some areas, tended to favor the maintenance of existing structures and values in the face of real and imagined threats to established traditions.

Toward the end of the war, Norwood became increasingly prominent on the national stage, and put forward his own ideas for the future of secondary education. These resonated well with the major reforms that were being developed by the president of the board of education, Herbert Fisher, who was appointed in 1916. Fisher's plans were to culminate at the end of the war in the Education Act of 1918. At this critical juncture, Norwood became established as a serious contributor to educational debate. At a high profile meeting in Oxford in August 1917, for example, Norwood took the opportunity in a speech on "Educational ideals" to present his credentials as a forward thinking reformer. He was generally sympathetic to the plans and role of the Board of Education, but he was clearly not afraid to be outspoken on some issues. He aligned his ideals explicitly with those of Plato, whose views, he declared, were "identical with the democratic ideals of the present day, i.e., universal primary and secondary education, State control, careful selection of the best talent, equality of opportunity." He also spoke in favor of the development of part-time continuation schools that Fisher was considering as part of his plans for reform. For secondary schools, Norwood suggested a broad framework for the curriculum and examinations. The curriculum should include Scripture, English, history, geography, mathematics, science, two languages, music, drawing, physical training, manual construction, and, in the case of boys, military training. Examinations would be based directly on the work of the school, with a first and a second school certificate to testify that the student had followed a course of study in an inspected school and had passed an examination. He emphasized that what he described as the "tyranny of the classics" in the curriculum should not be replaced by a similar domination by science. He also advocated that a national system should be established in which the Board of Education would be responsible for finance and inspection, while the LEAs would provide buildings and meals and attend to financial details. Teachers would be subject to public inspection, but they should have greater freedom than in the past. Finally, he called for the growth of a religious spirit through religious training.[2]

Such contributions made Norwood an obvious candidate for involvement in the Board of Education's reforms. On the other hand, his tendency to speak his mind in public may well have made board officials uneasy. When considering Norwood's potential appointment as a member of the SSEC, the response of the Board of Education officials to his potential role appeared lukewarm at best. The position arose because the newly created SSEC agreed to ask the board to provide for the inclusion of a head of a secondary school as an additional member, as well as representatives of the professions, commerce, and labor. The idea of these external representatives was turned down, but it was agreed to consider who might best fit the role of a secondary school head teacher. Accordingly, W.N. Bruce came up with a short list consisting of three names: George Smith, master of Dulwich College, C.H. Greene, headmaster of Berkhamsted Grammar School, and Cyril Norwood. Bruce described Norwood as "The Master of Marlborough College to which he went from Bristol Grammar School, which is on the grant list and where he was a great success. One of the three or four best headmasters in the country: was once in the Civil Service." However, despite these impressive qualifications he did not favor inviting Norwood: "I fear he may be too much engrossed with his reforms at Marlborough which are said to be extensive." Bruce argued instead in favor of Smith.[3] Another official demurred, pointing out that Smith was finding difficulties at Dulwich, and proposing that a headmaster of a school on the grant list would be most appropriate for such a position; Greene, therefore, appeared to him the most appropriate choice.

It was the president of the Board of Education, faced with this advice from his officials, who came down decisively in favor of Norwood. As Fisher pointed out: "Surely we ought not to miss this chance of securing Mr. Norwood if he is available? He strikes me as being the most interesting headmaster I have yet met."[4] Norwood was indeed available, and took up the post. Four years later, in 1921, he went on to become the chairman of the SSEC, and he remained in this position for a quarter of a century until he retired in somewhat different circumstances in 1946. His role in the SSEC undoubtedly gave him the opportunity to have a continuing involvement in national policy, an involvement that exceeded that of any other educationalist outside the Board of Education during this period. Yet he remained in some respects an outsider and found it difficult to allay the doubts of officials at the board as to his influence and judgment.

The Conservative Curriculum

In the postwar period, the Board of Education increasingly showed itself willing to offer great freedom to secondary schools. It encouraged them to develop stronger control over the curriculum, and to enhance the position of individual teachers. A number of commentators have noted that the board loosened some of the restrictions around elementary schools at this time, culminating in 1926 with the abolition of the compulsory elementary curriculum.[5] Developments in the secondary school curriculum have been less widely remarked. The reception given to these initiatives revealed the generally conservative nature of secondary school heads and teachers, and thus the difficulties that confronted any extensive reform of the secondary school curriculum.

There had been indications for several years that the Board of Education was willing to encourage secondary schools to take greater responsibility in the curriculum domain. Lord Eustace Percy, president of the board of education in the Conservative governments of the 1920s, claimed that "The whole tendency of the Board in recent years has been to give greater freedom in the curriculum of secondary schools."[6] He observed that time requirements for certain subjects had been relaxed as early as 1907, although a tradition of including particular subjects in the curriculum had survived. Circular 826, issued by the board in 1913, had recognized the need for greater elasticity of the curriculum.[7] At the same time, the board continued to offer advice to the schools on the curriculum, and indeed during and after the war it produced a succession of reports on the teaching of different subjects in secondary schools. These included natural science, English, history, modern languages, and classics.

Each of these reports remarked on the relative neglect of the subjects with which they were concerned, and the likely consequences for the nation if this were to continue. The report on natural science, "this Benjamin of subjects," pointed out that the public schools had tended to be lukewarm at best. Secondary education had recently expanded to new schools, but, the report suggested, "the older schools have not yet been entirely freed from all their prejudices, and the newer schools, in spite of their better balance of subjects, may perhaps have missed some of their opportunities."[8] According to this report, these weaknesses had been exposed by the war and its needs, but improving provision for science would also enhance prosperity in time of peace.[9] Similarly, a report on the position of modern languages, or "modern studies" as it preferred to describe them, was critical of their deficiencies and neglect. It noted that the war had revealed the full value of modern studies, as well as the dangers to the nation that would accrue if they did not develop further. Indifference and apathy among businessmen, politicians, administrators, and the public at large needed to be overcome if modern languages were to become more prominent in the educational system, and this would require an association with high ideals of the kind that classical studies had previously attained.[10] Again, it was in the secondary schools that further progress was most keenly sought, on the ground that it was here that the foundations would be laid for subsequent work.[11]

The report on classics, published in 1921, was of particular interest for the anxiety that it expressed about the future.[12] It was produced by a specially appointed committee that included a number of well-known and influential classicists such as

R.W. Livingstone, Gilbert Murray, Cyril Norwood, and Alfred Whitehead. The report celebrated the potential of classical studies in Latin and Greek for their educational contribution, not only for the few students who would examine them in depth, but for a large number of pupils who might benefit from at least an introductory course. It also emphasized that the classics comprised the foundations of modern European civilization and the source of contemporary ideas and principles, amounting to a unique cultural heritage that deserved to be sustained and preserved into the future: "By the study of the Classics we mean the study of these writings and monuments, and therewith of the languages and literature and art of ancient Greece and Rome, both as interpreting that civilisation and as being in themselves lofty and unique expressions of the spirit of man."[13]

Nevertheless, according to this report, such studies were in grave peril. Since 1900, it observed, they had gone into decline, especially Greek studies, and it insisted that this situation presented a danger that "the greater part of the educated men and women of the nation will necessarily grow up in ignorance of the foundations on which European society is built."[14] It urged that this trend should be resisted, and priority given to encouraging its study in different kinds of schools, universities, and other educational institutions. Finally the report emphasized its dire forebodings of "national disaster" if classical studies were not preserved, and if the cultural foundations of modern civilization were undermined as a result:

> No one who has given serious attention to the matter can doubt that the economic, political, social and moral welfare of the community depend mainly on the development of a national system of education which, while securing for every child in the country the equipment necessary for playing his part amid the complex conditions of modern society, will also provide his leisure with ennobling occupation and his life with a spiritual ideal. And we would submit that in such an education the study of the literature, art, science, history and philosophy of Greece and Rome cannot be replaced by any other which in both respects is so comprehensive and so effectual.[15]

This report thus provided eloquent testimony to the insecurities that surrounded secondary education, in this case, the fabric of civilization itself that might be vulnerable to attack from within and without.

Notwithstanding such high-level advocacy for more thorough treatment of the various major subjects, the board also recognized that secondary schools needed to be in a better position to resist pressure that might lead to the overcrowding of the curriculum. It was this that provoked it to revise Circular 826 with a new set of recommendations, Circular 1294, issued in December 1922. The permanent secretary at the board, E.H. Pelham, noted that the main purpose of the new Circular was "to emphasise the fact that the growing demands of different subjects are tending to overcrowd the curriculum, even if they are not impossible to reconcile."[16] He continued, "We want to get schools frankly to recognise the position, and to seek relief by going further than has hitherto been usual in the direction of reducing the time allocated to different subjects, and in omitting some subjects from the curriculum altogether, at least for some pupils and some stages of the course."[17] However, he acknowledged that this also raised awkward issues about the board's own advice to schools in relation to the curriculum. For example, he pointed out, it was difficult for the board to press for compulsory Latin in secondary schools: "Any pressure by the Board in favour of the more general study of this one subject appears to be hardly consistent with the greater freedom in arranging the curriculum which is the main concern of the Circular to offer to the schools."[18]

Circular 1294 duly emphasized what it described as an increasingly acute problem, the "squeeze of subjects" in the secondary school curriculum. In the light of this, it made a strongly worded call for schools, "in planning their curricula, to exercise a greater freedom than has hitherto been the case."[19] It therefore welcomed proposals from schools for lightening their curricula.[20] Yet it was soon apparent that there was very little response forthcoming from schools, and a year later Pelham decided to elicit the views of inspectors around the country on the reasons for this apparent lack of interest.[21] The major factors indicated in this survey were the stultifying effects of examinations and the conservative tendencies of teachers and head teachers in secondary schools.

The examinations system imposed a strong influence over the secondary school curriculum in the 1920s and 1930s, and this was already widely recognized in the inspectors' responses to Pelham's inquiry.[22] One district inspector observed that "Inspectors are unanimous that the controlling check on experiment is the First School Exam. I must confess to being uneasy at the tremendous hold this exam is getting over the Schools, and the tendency of LEAs to estimate Schools according to exam results."[23] Another inspector noted: "Schools are controlled now rather by the First School Examination than by regulations. They cannot experiment below this stage without risking the candidates' chances of obtaining credit or a *matriculation* certificate and most headmasters regard the risk as too great."[24] Another HMI, F.R.G. Duckworth, who was later to become senior chief inspector, also commented pungently on what he called the "fear of examinations" in the Middlesex area: "The worship of examinations flourishes in the country and the LEA's habit of publishing results fosters this."[25]

The constraining influence of examinations was exacerbated by the habitual caution that prevailed in most secondary schools. According to one inspector, head teachers were not anxious for freedom and often did not grasp the significance of Board of Education circulars. Moreover,

> Only an exceptionally strong man or woman could stand against the forces that support the present curriculum, tradition, public opinion, economy, vested interests, previous training, inertia, and the desire for a peaceful life. The teachers are recruited naturally from the unenterprising section of the community, particularly in the case of men and are trained at school and university tamely to accept the opinions of others. Moreover to most Head Masters the ideal product is the examinational prodigy who will do brilliantly at the University. The other pupils must follow haltingly in this track.[26]

Another pointed out that teachers and LEAs were currently distracted by other issues, teachers with salary questions and LEAs with economy, with the result that "neither goes out of the way to tackle this special aspect of the curriculum which is regarded as especially thorny."[27] This theme was pursued further by F.B. Stead, who argued: "The Circular was not issued at a very fortunate moment. It is difficult to get the teaching profession to attend to more than one or two things at a time: and in the last year or two the attention of the profession has been largely engaged on such matters as the Burnham scale and its application, and the meaning of 'full-time service.' " Moreover, according to Stead, "Even at the best of times Circulars are apt to be read and then laid aside and forgotten . . . . As things are it may easily happen that even an important Circular will be stillborn."[28] Such factors inhibited reform and experiment in secondary education. At the same time, it was evident that there was little agitation for further changes to the Regulations for Secondary Schools, and their lack of elasticity was rarely regarded in the schools as a problem.[29]

Gender Differentiation of the Curriculum

Concern about overcrowding of the curriculum and pressure to allow schools increased freedom in this area was especially acute in relation to the secondary education of girls. This was articulated in a report produced by the Board of Education's new consultative committee at the end of 1922, a report that examined the differentiation of the curriculum for girls as distinct from boys. Women head teachers were divided on this issue and were often hesitant to go in the direction of greater curricular differentiation especially as it might lead to reduced status for their schools. However, at this time, the idea of increased differentiation of the curriculum along gender lines was taken up with enthusiasm by Cyril Norwood. Although he had never been himself directly involved in the education of girls, Norwood emerged as a champion of an ideal of secondary education that took account of the future domestic roles of the majority of women.

The consultative committee's report on the differentiation of the curriculum for boys and girls emphasized the physical and mental differences between the two genders and also pointed out that they had distinct social functions. It was the general conception of the social functions of men and women, it argued, that should primarily determine the education of boys and girls.[30] It suggested that the earlier forms of education for girls and women, until the mid-nineteenth century, had recognized such differences and had done so because women were regarded as the weaker sex. This had been followed by a period in which the education of girls was sought to be made equal to that of boys, and this was done by making it as far as possible the same as the education of boys. It argued that girls' education was now ready to enter a third stage in its development, in which it would be differentiated from that of boys but in a spirit of equality between the sexes. After all, it reasoned, "Dissimilars are not necessarily unequals; and it is possible to conceive an equality of the sexes which is all the truer and richer because it is founded on natural recognition of differences and the equal cultivation of different capacities."[31] Nevertheless, it concluded, such differences were not so great as to prescribe one curriculum for boys and another for girls. It preferred to envisage a relaxation of requirements and an increase of freedom of choice for the period of studies leading up to the First and Second School Examinations, and it expressed confidence that such freedom would usher in "a time of progressive experiment, in which teachers will seek with vision and with courage to provide the course and use the methods which will best suit the capacities and the tastes of their pupils."[32]

The committee pointed out what it regarded as a clear improvement in the provision of secondary education for girls in England and Wales since the passage of the Education Act of 1902. From 99 girls' schools being included on the grant list between 1904 and 1905, there were 450 such schools included between 1921and 1922. This was in addition to an increase in the numbers of coeducational schools on the grant list over this time from 184 to 331. Overall, the number of girl pupils in schools on the grant list had increased more than fivefold in less than twenty years from 33,519 to 176,207, while the number of boy pupils in such schools had tripled in the same period from 61,179 to 184,408.[33] Under the existing Regulations for Secondary Schools, there was compulsory provision in girls' schools of practical instruction in domestic subjects such as needlework, cookery, laundry work, housekeeping, and household hygiene, which were generally substituted with woodwork and metalwork in boys' schools. The committee noted a wide range of criticisms of the curriculum provided for girls, for example, that it was too academic, overburdened, too rigid, too similar to the curriculum for boys, and that it might lead to "overstrain." It acknowledged that girls needed to be prepared not only to earn their own living and to be useful citizens, but also (and unlike boys) to be makers of homes.[34] This should be achieved by providing greater elasticity for the curriculum.

There were some headmistresses who were keen to take advantage of the greater freedom recommended by the consultative committee's report. For example, the experienced Sara Burstall, head of Manchester High School for Girls and president of the Association of Head Mistresses, was highly critical of what she regarded as the restrictive regulations on the curriculum. She suggested that science, French, and mathematics should not be required for some girls in their advanced courses, emphasizing that "we want more freedom to teach our girls fewer subjects at a time, and to drop some out altogether if we think fit for individual pupils."[35] According to Burstall, rigidity in the curriculum "can and does harm educationally, and too much interference and regulation destroys initiative and joy in the teacher." She objected to undue interference form the board officials, and championed the need for the "average girl who will be a wife and mother" to be allowed an education that would fit them well for "home life."[36]

Nevertheless, many other headmistresses were loath to modify the curriculum in this way. Such reluctance was regarded by inspectors as a general unwillingness on the part of girls' schools to develop an inferior version of secondary education. One inspector claimed that "Head Mistresses will not take the lead because they look on any diminution of the claims upon girls as compared with boys as tantamount to an admission of sexual inferiority."[37] This view was strongly echoed in a summary of inspectors' responses to Pelham's original enquiry into the impact of Circular 1294. According to this summary,

> A rather curious attitude on the part of Headmistresses of Girls' Schools is disclosed. Headmistresses, it is said, fear to drop subjects which are taken in boys' schools as to do so might be interpreted as a sign of "sex inferiority"; they prefer a crowded curriculum, and subjects, especially those of a literary, artistic, and musical character, are introduced by curtailing the time available for other subjects like Science and Mathematics.[38]

Thus, there was evidence in girls' secondary schools no less than in those of boys that heads and teachers were generally not prepared to make use of the increased curricular freedom being offered by the board. As a means of countering this attitude, one inspector suggested that a committee might be established to investigate the curriculum and the principles that should underlie its construction.[39] Pelham was unconvinced by this notion and concluded that the schools should be left to go on "working quietly" and "gradually evolving." His distaste for the idea of a committee was revealed in his final comment: "A Committee would be a sounding board for the enthusiasts!"[40]

At the same time, there was continued pressure from the Association of Head Mistresses to recognize the nonacademic subjects in the First School Examination. This could be achieved, according to the association, by amending the examination system and Secondary School Regulations to allow a wider range of nonacademic subjects to be taken.[41] Board officials and chief inspectors were doubtful about this course of action because it threatened the character and standards of secondary education, although they were happy to concede the argument about gender differences necessitating further differentiation of the curriculum. At a special meeting held on the issue in November 1927, the parliamentary secretary to the board pointed out that "what they had to remember in connexion with the education of girls was its dual object; it must prepare not only for a great many careers and professions but for marriage and home life." It was this consideration, "together with the greater liability of girls to overstrain," that in her view made it necessary to create a greater degree of elasticity in their curriculum.[42] Yet there were definite limits to this elasticity if it meant undermining the tenets of secondary education as a whole. The board's permanent secretary insisted that the study of foreign language must be retained for all secondary school pupils, and it should indeed be regarded as "the distinctive feature of secondary education." The chief inspector, F.B. Stead, "doubted the reality of the distinction implied by the terms 'Academic' and 'Nonacademic' ability and he did not believe that the average pupil from a Secondary School which was reasonably well-staffed found the First School Examination a serious obstacle." The meeting ended without clear resolution, but it was agreed to obtain further advice from headmasters, and also "to inform Dr Norwood of the action which the Board was taking."[43]

The reference to Dr. Norwood was significant. Not only was Norwood the chairman of the SSEC, but he was also a public enthusiast for reform who was particularly active in advocating further moves toward a differentiation of the curriculum for girls as distinct from boys. In a series of articles and public speeches, he made his views on the matter very clear at every opportunity in the 1920s and early 1930s. In October 1928, for example, he affirmed emphatically in an interview with the newspaper the *Daily Mail* that "girls should *not* be educated like boys." Perhaps, he recognized, there were individual girls who might be capable, "intellectually, and even physically," of the same work as boys. Nevertheless, he insisted,

> When you consider the 100,000 or so of girls of 12 to 18 who are now being educated in the secondary and public schools—the number has increased two and three times since the war—is it not very short-sighted to suppose that a stereotyped course of learning will suit all of them? The majority will eventually marry. At school they are taught exactly as if they were going on to university.[44]

Such intransigent opinions met with some resistance. Norwood was invited to open Westonbirt, a new public school for girls, in 1928, and took this opportunity to express at the opening ceremony his views on women in general and on girls' education in particular. He emphasized that girls' education should be neither unduly academic in its outlook nor too closely modeled on that of boys. The magazine *Time and Tide* responded acidly: "To-day it is only here and there that we find lingering the point of view of which Dr Norwood is so distinguished an exponent."[45]

It is notable nonetheless that Norwood continued to receive invitations to speak at girls' schools. Three years later, he returned to Westonbirt school on the occasion of a visit by Princess Mary, this time to pontificate on the "things girls do better than boys." Boys, he proposed, were better than girls at advanced mathematics, chemistry, Rugby football, and boxing, while girls were better at English literature, winning the Newdigate prize at Oxford, probably swimming, and "having a conscience." He noted that some women complained about their position and about the wrongs of men and how they had been kept down for generations by man-made laws: "He had heard his wife use that argument (laughter)." However, he reasoned, this view was merely a fallacy. The most urgent need was for cooperation, not rivalry, and for this reason the education of women should recognize that most of them would be married one day: "Most of the women in the country did get married, and that meant running a home; and he could not but think that the ordinary education of a girl should always keep that in view." All single girls should be trained in a professional occupation, so that they could be independent if necessary, but by the same token the education of the majority should be geared to the home. In the future, he suggested, there was likely to be a woman prime minister and a woman archbishop of Canterbury, "But what he wanted to say in all seriousness, was that the best thing that women could do in the coming generation was to safeguard that family life, of which their own Royal family was so striking an example (applause)." Thus, he concluded,

> He would like to say that the two finest words in their language were "home" and "mother," and they could not think of the one without thinking of the other at the same time. The country had got to fight hard in the coming generation for its world position and for its standard of life, and in that struggle he knew that women were going to play a great part in business and in the professions, but he believed their greatest part was going to be in the home.[46]

The punctuations of laughter and applause added in the newspaper report of the speech suggest that this at least was a receptive and willing audience for these frankly old-fashioned views.

Norwood's interest in this topic was at its most conspicuous in the late 1920s and early 1930s, and this is explicable in terms of the threat that women appeared to pose to the established social order at this time. The days of the suffragettes may have passed, but now there were new dangers represented in the political advances being made and the subversive ideas of many middle-class women. In the area of secondary education, the ambitions of women seemed, to conservatives such as Norwood, to pose a threat to the integrity of society in relation to the primacy of home and family. The number of girls involved in secondary education also appeared increasingly at odds with Norwood's conceptions of the character and ideal of this kind of education. The outcome was that Norwood resorted to emphasizing the traditions of English education and society as a means of warding off the perils of the future. This had become highly characteristic of his response to change. It was a response that was once again rooted in social anxiety, lest new ideas and interests serve to undermine the civilization and values to which Norwood was attached.

Norwood's references to tradition had also become a convenient means of differentiating between different groups in society. In this case, gender provided a basis for emphasizing the differences between two kinds of mind, the male and the female. This represented a basic typology that affirmed natural and historical differences between men and women. At the same time, there were other emerging social threats that also attracted his displeasure. These also led him to idealize a particular kind of tradition associated with secondary education, and to develop another typology that would serve to justify its continuance.

The Next Steps?

During the 1920s, growing economic and industrial problems led to a period of retrenchment in education that undermined the ambitions of the Education Act of 1918.[47] In this context, the Labour Party, emerging as the principal opposition party to the Conservatives in British politics and seeking to offer a radical alternative, argued in favor of extending secondary education to all children. The key expression of this policy was the report *Secondary Education for All*, produced by the influential economic historian R.H. Tawney. This work declared that secondary education should be regarded as the right of all, "secondary education being the education of the adolescent and primary education being preparatory thereto."[48] The division of education into elementary and secondary was, it insisted, "educationally unsound and socially obnoxious."[49] On the other hand, it acknowledged that there should be more than one type of secondary school, and that "local initiative and experiment" should be encouraged. Thus, it continued,

> There is no probability that what suits Lancashire or the West Riding will appeal equally to London or Gloucestershire or Cornwall, and if education is to be an inspiration, not a machine, it must reflect the varying social traditions, and moral atmosphere, and economic conditions of different localities. And within the secondary system of each there must be more than one type of school.[50]

This broad approach suggested that secondary education should be defined not in terms of a particular kind of curriculum or ethos, still less as an expression of social distinctions, but simply in relation to the age range of pupils.

This reappraisal of the nature and meaning of secondary education was continued in 1926 through the publication of the report *The Education of the Adolescent* by the Board of Education's consultative committee. Commonly known as the Hadow

Report, after the chairman of the committee Sir Henry Hadow, it accepted that secondary education was being interpreted in an unduly narrow sense. It proposed instead that the term "elementary" should be abolished in favor of the word "primary," and that the name secondary should be given to "the period of education which follows upon it."[51] The schools that currently were described as secondary schools would be called by the name of grammar schools, while other, newer forms of secondary school would be called modern schools. Thus, it concluded, "On such a scheme there will be two main kinds of education—primary and secondary; and the latter of these two kinds will fall into two main groups—that of the grammar school type, and that of the type of the modern school."[52] It conceded that such a change would involve a fundamental shift in attitudes and expectations: "We admit that we are here walking on difficult ground, and that there are fires burning beneath the thin crust on which we tread."[53] Nevertheless, it was clear as to the practical outcome of what it was proposing:

> It is that between the age of eleven and (if possible) that of fifteen, all of the children of the country who do not go forward to "secondary education" in the present and narrow sense of the word, should go forward none the less to what is, in our view, a form of secondary education, in the truer and broader sense of the word, and after spending the first years of their school life in a primary school should spend the last three or four in a well-equipped and well-staffed modern school (or senior department), under the stimulus of practical work and realistic studies, and yet, at the same time, in the free and broad air of a general and humane education, which, if it remembers handwork, does not forget music, and, if it cherishes natural science, fosters also linguistic and literary studies.[54]

On this view, then, secondary education would be defined simply in terms of education that is appropriate to a particular age group, but there would be different types of secondary education catering to different kinds of needs and orientations.

As the chairman of the SSEC, Norwood was in a key position to negotiate the implications of these proposals. He did so in a forthright and aggressive manner that challenged the expectations of a number of interested groups. The focus of his attention was the School Certificate examination, which continued to be debated earnestly by headmistresses who wished a looser arrangement that would permit different subjects to be combined. In the wake of the Hadow Report, the School Certificate examination also raised issues about the nature of the provision for modern school, as distinct from that of grammar school pupils. At the heart of Norwood's concerns was the established character of secondary education, which he set out to defend by questioning the position of nonacademic pupils, especially girls.

The nature of Norwood's position was made clear in a series of exchanges at the end of 1927 and the start of 1928. In October 1927, a deputation from the Head Mistresses' Association presented the SSEC with a proposal that the subjects to be taken in the School Certificate should be less restricted than it was in the current arrangements in which the core subjects of mathematics, science, and French were compulsory. The suggestion was referred for further consideration, to return to the council for a decision the following February. Norwood immediately raised strong concerns with Maurice Holmes, the permanent secretary at the board:

> It is obvious that this proposal (if approved) will profoundly affect the content of secondary education, since it will be possible to omit Mathematics, Science, or French entirely. It is probable that French is the subject that the Headmistresses are most likely to omit.[55]

He complained also to the chief inspector, F.B. Stead, again arguing that the proposal of the headmistresses would "wreck the whole elaborate system." The only way in which it would be possible to proceed further with their ideas, he argued, was by having them "willing to accept another Certificate altogether." In the meantime, he insisted, their proposition "that things [i.e., subjects] which are not equal to one another shall be unequal to the same thing" was unacceptable.[56]

Norwood, who was by now committed in his opposition to the headmistresses' plans, set about elaborating on his rationale and proposing an alternative in a lengthy memorandum to Stead at the beginning of 1928. He linked his current concern about the School Certificate examination to the increase in the number of secondary schools since the Education Act of 1902 and suggested that it demonstrated that some of this expanded clientele were not fully suited to a full secondary education as it was currently understood. In the circumstances, the only alternatives were to dilute the character of secondary education for all pupils, or to maintain it in its established form and find a different type of provision for those pupils for whom this would be more appropriate. He reminded Stead that whereas in the years 1904 to 1905 the number of pupils in schools from which the board received figures was 33,519 girls and 61,179 boys, these figures had grown in twenty years to reach 173,273 girls and 194,291 boys. In Norwood's view, this was a "veritable flood of new material," and it was "not of the best intellectual quality." In addition, he predicted further growth that would arise from the four-year courses of central schools and as a result of the recommendations of the Hadow Report. If the School Certificate examination were revised in such a way as to allow all of this enlarged group to be able to obtain it, the tradition of secondary education itself would be undermined: "In accordance with the law that the worse drives out the better currency it is probable that in twenty years the conception of Secondary Education, as established by the Board, would disappear."[57]

Norwood proposed therefore that since most girls went on to marry, their secondary education should reflect to a greater extent their "probable future vocation," and that the "great, and increasing" number of boys who were in academic difficulties should also be given a different type of examination. Thus the School Certificate examination should be maintained as it was for those pupils who were suited to it, while a new examination, which he described as a "General Secondary Certificate," should be invented to meet the needs of other pupils. This new set of arrangements would soon establish itself in use; and according to Norwood, this would be far preferable to the alternative, "to widen the School Certificate so that it would cover everything," which would "seriously throw back the great work of establishing a sound tradition of Secondary Education that has been accomplished in the last twenty-five years."[58]

The views expressed by the chairman of the SSEC naturally won a measure of sympathy from senior board officials, but at the same time the board was keen to exercise caution. Stead summarized the role of the School Certificate examination in relation to the established ideal of secondary education. It was, he noted, intended to be a mass examination that pupils of "ordinary ability" who had followed the secondary school curriculum should be able to "take in their stride" at the appropriate ages. It was supposed to provide a test of the course of general education pursued by pupils, a test that should include a modest level of performance in English subjects, a foreign language, and mathematics or science. He concurred with Norwood that it was based on the established tradition of secondary education in England: "It was expressly designed to correspond to the prevailing conception of the content of a secondary education as embodied in the Board's Regulations, a conception which is of course much older than the Regulations or the Board."[59]

Nonetheless, the pragmatic Stead thought it possible to develop minor concessions that would go some way toward meeting the concerns of the headmistresses, "without destroying the general structure of the examination and damaging the educational work of schools properly described as Secondary." This would involve tinkering with the subject requirements that were required for a pass. He put this suggestion privately to Norwood. The latter replied that it would "do for ten years," and explained that "he was thinking of the forthcoming influx of pupils into modern schools." His idea of a General Secondary Certificate was in fact aimed not only at nonacademic girls but also at "the needs of the pupils pursuing postprimary education as set out in the Hadow Report." Stead was not impressed, pointing out to Norwood that this issue was not pressing at this stage and could be dealt with separately. However, Norwood's response reflected his longer-term strategic approach to the state of secondary education as a whole, which he insisted was in need of decisive action to maintain its established tradition.[60]

Subsequent developments revealed Norwood's exposed position. At a meeting to discuss the SSEC's response to the headmistresses, held in March 1928, Norwood put forward his proposal of maintaining the current School Certificate while also introducing a General Secondary Certificate with looser subject requirements.[61] This action was greeted with alarm at the Board of Education. Maurice Holmes felt that Norwood was acting "rather strangely," and it was not the right time to go forward in this way. He suggested that Norwood should be invited to the board to discuss the situation informally, while making clear the general sympathy of the board with his views: "On the main question of the inviolability of the First School Examination his views are eminently sound and we do not want to antagonise him."[62] Another official, W.R. Richardson, was more forthright in his opinion of Norwood, who despite his position and background was beginning to appear unpredictable, if not unreliable. On the one hand, Richardson conceded, "We have to recognise that Dr Norwood is in a difficult position, and that it is necessary to deal with him in a friendly and unofficial way." On the other, he argued, Norwood's proposal of a General Secondary Certificate that could be used as a general examination for modern schools was "lamentable." Establishing a suitable examination for modern schools was an issue that was "at once delicate and important" and would require "all the prestige of a fresh and living attempt to focus the work of the new schools" if it was to appeal to teachers and employers. And yet, Richardson continued,

> In the midst of our rather anxious deliberations on this point Dr Norwood presents us, rather contemptuously, with an Examination not good enough for the ordinary Secondary School pupil, but good enough for those who cannot take "the full Secondary School course," and "are not of the best intellectual quality," "not the right sort of material for the First School Examination."[63]

This, he contended, would endanger the whole policy of the Board of Education. It was also "extremely short-sighted" from the point of view of the secondary schools, as the alternative certificate that was being proposed would be too easy for most pupils of senior schools. Neither would it be appropriate for schools that were seeking to develop a special industrial or commercial approach: "It is not only probably low in standard, but inappropriate in content."[64]

In the face of this kind of opposition from his natural allies in the board, Norwood beat a hasty tactical retreat. At the subsequent meeting of the SSEC, Norwood withdrew his proposal of a General School Certificate, "as the result of communications he had received from the Board."[65] Nevertheless, he was unrepentant and continued to put forward similar ideas in public forums. One such occasion was his presidential address to the educational science section of the British Association for the Advancement of Science, presented in Glasgow in September 1928. Under the bold title "Education: The Next steps," this high-profile speech put forward a set of radical proposals for further educational reforms. He put these in the context of a growing national trend that he identified as a belief in the value of education. Despite the resistance of an unrepresentative minority, he opined, the nation as a whole was now "unwilling to think of a large mass of its members as merely raw material to be utilised in its course from the school to the scrap-heap; it believes that each boy and girl has a right to be trained as an individual."[66] According to Norwood, this meant in turn that the national system of education should now be completed in order to "provide for all the varying and complicated needs of a great nation of the twentieth century."[67] Such a development would entail implementing the proposals of the Hadow Report of 1926, with the aim of establishing primary education for all upto the age of 11-plus, followed by secondary education for the majority of pupils who were 15 or older, for many who were 16-plus, and for some who were 18-plus: "It means as an ideal that all children would go forward after eleven on parallel lines, following the course best suited to each."[68] He agreed with the authors of the Hadow report that the creation of a system of different types of secondary schools would not hamper or cripple the secondary schools that already existed. The new schools, he suggested, would begin with their 11-year-old pupils in much the same way as the established schools would, but they "will always seek to develop the hand and the eye, and in their last two years will develop a practical bias."[69]

In addition to sketching out a system of secondary education for the future, Norwood also returned unabashed to his contentions about recent developments. He pointed out that while the content of secondary education had not changed, and it remained academic in spirit and outlook, the number of schools had more than doubled, and the number of pupils had increased by more than four times. It was this, he declared, that had led to difficulties, including the overstrain of many pupils (especially girls) obliged to struggle with an unsuitable curriculum, and a loss of freedom for teachers hampered by examinations. According to Norwood, this should not mean that the character of secondary education should be changed. Far from it, "The standard of secondary education in England is high, and is something of which we have a right to be proud. Its methods and objects are the fruit of long experience and of the efforts of several generations."[70] Nor should the School Certificate and Higher School Certificate be tampered with. Rather, he proposed, there should be two kinds of certificate: "one which shall fulfill the academic conditions and maintain unlowered the existing system which causes no difficulty to the boy or girl of average academic ability, and the other which shall be a proof that the boy or girl has taken at school that course of education which in the particular case was the most fitted."[71] This would also encourage greater variety within secondary education. Moreover, Norwood declared for good measure, at some stage in the future it would be necessary to challenge the entrenched position of external examinations in secondary education: "I believe, therefore, though the time is not yet, that the right course will be to abolish all external examination for the average boy and girl, though leaving it as the avenue to the universities and the professions."[72]

It was hardly surprising that Norwood's predilection for public discussion of major issues created a high state of nervousness among Board of Education officials. In this instance, he clearly remained committed to a set of changes on which he had recently been carefully but definitely rebuffed. More broadly, these discussions demonstrated clearly a number of basic tensions around the ideal of secondary education at a crucial stage in its development. There was in the first place a strong emphasis on established tradition, including anxiety and fear at the prospect of undermining or losing it. Such fears were expressed most forcibly by Norwood who was determined to act decisively to protect it, and they were shared by board officials who were nevertheless more circumspect and subtle in their approach. To Norwood, defending the tradition meant being rigid about the established ideal and enforcing divisions to exclude those who did not conform or meet the required standard. Educational division also spelled social division, in this case along both gender and social class lines. He was vague about what provision would be appropriate for those who were excluded from secondary education of the academic type.

Norwood was at the peak of his influence on educational policy at a national level, and yet somehow he remained an outsider. He was widely respected as a formidable exponent of ideas and arguments, but he was also felt to be unreliable and unpredictable among the policy elite; too prone to expressing his view in public speeches and even in newspapers, too fond of attracting attention to his ideas, too open and frank in his opinions. Officials and politicians much preferred discreet discussions in private. He had been frustrated in his key objective of defending the ideal of secondary education in the policy arena. In his own professional domain, he had succeeded in gaining access to the citadel of the elite boarding schools that symbolized this ideal in its purest form. Yet even here, he was to encounter opposition in different forms that threatened to undermine the ideals that he championed, and it is to this aspect of his work that we now turn.

Chapter 6

Marlborough and Harrow

While Cyril Norwood was cultivating a national reputation in education policy, he was also engaged in promoting the fortunes of two leading public boarding schools. As the master of Marlborough College in Wiltshire from 1917 until 1926, he set about emphasizing the traditions of a prominent school that was still comparatively recent in its origin. A more difficult assignment awaited him as head of Harrow School from 1926 until 1934, a very well established and prestigious public school, but one that was beset with internal disputes and factions. In a sense, both schools provided a retreat from the national policy debates in which Norwood had become immersed. Yet there were also significant connections between Norwood's professional life as a headmaster, and his public role in education policy. In setting out to reform and modernize these major public schools he was attempting to find common ground between the independent sector and the newly established state system. The opportunity to appreciate at close hand the traditions with which Marlborough and Harrow were associated also allowed him to proselytize these ideals more broadly. At the same time, Norwood encountered difficulties at Marlborough and especially at Harrow that demonstrated in vivid fashion the contested character of these traditions, and the social divisions that continued to exist between public and grammar schools.

Summoned by Bells

Marlborough College was founded in 1843, the year after the death of Thomas Arnold, mainly for the sons of clergymen. Under its original scheme, at least two-thirds of pupils were to be the sons of clergymen of the Established Church, who would pay 30 guineas per annum, while not more than one-third would be the sons of laymen, who would pay 50 guineas per annum.[1] Its principal aim was therefore to provide "a first-class education for the sons of clergymen," unlike other public boarding schools that were beyond the means of the majority of the clergy.[2] It could also boast a strong Arnoldian influence, as its major headmasters during the nineteenth century included G.E. Cotton, who had been on Thomas Arnold's staff at Rugby, and G.G. Bradley, who had been a Rugby housemaster. It was not included within the scope of the Clarendon Report of the 1860s. However, the Clarendon commissioners invited it, along with Cheltenham College and Wellington College, similarly of recent vintage, to provide information about its organization and results.[3] Thus, although Marlborough was by far the most recently founded of all the schools with which Norwood was associated as pupil, master, or headmaster, it had much to appeal to him. Its main clientele included the sons of clergymen such as himself, there was the association with Thomas Arnold, and—although it was a boarding rather than a day school—it shared with Merchant Taylors the aim of providing less expensive education than that of most elite public schools. Moreover, as its pupils were from a scholarly background and aspired to individual improvement, it tended to favor an academic curriculum not completely distant from those of the grammar schools of Leeds and Bristol.

The educational career of Frank Fletcher, master of Marlborough College from 1903 until 1911, is instructive in relation to the parameters established for such positions at this time. Looking back on his life and career in 1937, Fletcher noted that he had spent most of his academic life in English public schools: seven years as a boy at Rossall, nine as an assistant master at Rugby, eight at Marlborough, and then the next 24 years as the headmaster of Charterhouse School. He had read for Greats at Balliol College, Oxford, with R.L. Nettleship as a tutor, and with Benjamin Jowett as the master of the college. After two years as a classical tutor at Balliol, he went on to become a tutor at Rugby School before being appointed as master of Marlborough College. His appointment at Marlborough had been in spite of his being a layman, although he was helped in this, firstly, by his "happily married" status, and secondly, by his being a classical scholar. The chairman of Marlborough's council, John Wordsworth, the Bishop of Salisbury, would have preferred a clerical appointment, according to Fletcher's account, and the appointment of a layman apparently caused much surprise, but "from that time the clerical limitation was definitely broken down, and to-day there is, so far as I know, no public school to which a layman has not been appointed."[4]

The bishop of Salisbury had continued to raise difficulties following Fletcher's appointment as master. In particular, Fletcher remembered, "He was dilatory in giving his consent to my preaching in chapel, and then, after satisfying himself that I had the requisite knowledge, he gave me formal leave in a legal-looking document, for which I had to pay a fee, and he made three conditions, that I should not preach from the pulpit, or in a surplice, or at the Service of Holy Communion."[5] Fletcher was not embarrassed by these restrictions to his role, but he used them to clarify his position as a headmaster by dressing in gown and hood and addressing his pupils from a lectern. Overall, he saw the free admission of laymen to headmasterships over this period as a relief and encouragement to the profession, allowing "a man who feels a vocation to schoolmastering" to be "no longer forced to ask himself whether he has also a call to ordination, knowing that, if he has not, the chances of the best work will be denied him."[6] Norwood was in a similar position to Fletcher in a number of respects, being married, a classical scholar, and a layman, and so to some extent at least his path to becoming master of Marlborough College had already been eased.

Norwood's application to Marlborough College naturally made much of his successes at Bristol Grammar School, but it also addressed the evident differences between Bristol Grammar and schools such as Marlborough. He was careful to emphasize that he valued freedom and the inculcation of character in his approach to teaching, and to relate this explicitly to the established approach of Marlborough:

> Though this is not the place for me to set forth my views on education, I may perhaps say that I have hitherto followed the method of appealing to the initiative of the boys under my charge, and trusting much to the good results of a carefully—observed freedom; this method I must follow, since I do not believe that any other is of equal value in the building of character and the training of citizenship which at Marlborough must be the first considerations.

He acknowledged also that his experience was in urban day schools rather than in boarding schools such as Marlborough, but he argued that this might be an advantage rather than a difficulty, on the grounds that "in the changes which must inevitably follow the end of the war it may be of service for the Headmaster of a great public school to have been in close touch with the life of the great cities and modern industrial conditions, and to know all that vast and growing side of national education which is represented by the modern Universities and Whitehall."[7] Thus he sought to accommodate social differences and his sympathies for an encroaching State with the elite position and independence of Marlborough College. He was duly appointed on this basis, but there remained many who failed to be convinced.

From the beginning, Norwood emphasized the importance of promoting the distinctive tradition of Marlborough College. This priority was given particular weight because of World War I and the heavy casualties suffered by former pupils and masters of Marlborough and other public schools, which led to calls for suitable forms of memorial to be established. At his first general Marlburian Meeting, in May 1917, he made it clear that while he was not "a member of your race by blood," he felt he was one "already by adoption" and hoped that "while I am Master the School will fill as high a place in the national education as it has done hitherto."[8] In this spirit, he argued forcefully in favor of "a visible memorial and an outward sign" to mark "the greatest and finest cause that Marlburian boys have ever fought in," and "the greatest sacrifice the School has ever made, or please God, ever will make."[9] Scholarships for the orphans of Old Marlburians who had fallen in the war would not meet this purpose, he urged. The school was especially in need of increased laboratory accommodation, but it would be "unthinkable" to establish a laboratory as a "memorial to the dead." A new music room or a new library would, he suggested, only cater to a restricted group and so would also be unsuitable as a memorial. He proposed, finally, that a fitting memorial should be associated with the Chapel of Marlborough College—"that glorious Chapel which is the centre of Marlborough's very life"—in the form of a cloister. This, he insisted, would provide the most permanent satisfaction and inspiration, "since day by day the boys will be along this Cloister, this corridor, which to them will be writ full of heroic names and its appeal will be all the stronger in their lives just because it is silent."[10]

Memorials continued to be a focus of keen interest at Marlborough College. At the annual Prize Day held in June 1921, Norwood was able to welcome a series of gifts from the mother of an Old Marlburian who had been wounded and killed at Gallipoli during the war: a herbarium and a piece of land on which would be set a stone bearing the name of her son. He thanked the mother of another casualty of the war for her gift of £1,200 to found two scholarships confined to the sons of the clergy. He was also able to describe the development of the war memorial, which—as had been finally agreed—would comprise a hall alongside the west of the chapel.[11] This was completed and opened in 1925, with a single word as the inscription above the entrance doors—"Remember."[12]

The specific tradition of the school, as Norwood regularly defined it, was about public service. In one sermon that he delivered as master of the College in March 1921, he took as his text the Biblical passage of II. Thessalonians ii, 15: "Stand fast, and hold the traditions ye have been taught." He stressed the importance of spiritual as opposed to material values, declaring that "in all your work here and in all your life here you are learning the same great tradition, that it is the business of the public school man to serve his generation." According to Norwood, the Army and Navy, the Church, the medical profession and the nursing profession were noteworthy as professions that were generally entered "not in the hope of making money, but with the desire of rendering service." This was also in his view the tradition of the great school and what Marlburians should stand for, "bound together in a spiritual unity because at bottom they respond to the things of the spirit and recognise only spiritual values as real."[13] Social service was also encouraged, for example, through a camp in the college precincts with a party of boys from the town of Swindon during the summer holidays. The school magazine voiced somewhat exaggerated optimism in relation to this latter initiative, hoping that "If a few Marlborough boys in a spirit of friendly comradeship and without patronage entertained in their own school thirty or forty boys of a class who necessarily spend their days within the streets of an industrial town, . . . nothing but good might result."[14] It was this spirit of public service that the war memorial and other monuments to the school's tradition were intended to convey.

At the same time as he was promoting the traditions of Marlborough, Norwood was also concerned to bolster its academic reputation. As at Bristol Grammar School, this was achieved partly by giving attention to the teaching facilities. In November 1917, he made a formal request to the College Council to agree to a wholesale improvement of the buildings and teaching space. In particular, he regarded the science buildings as "wholly inadequate," on the grounds that "They have grown up piecemeal, and are in part very old-fashioned. The rooms are, in some cases, small, and ill-arranged. They are barely sufficient for the small amount of science work which is done at present, and they are hidden away as if the College were half-ashamed of them."[15] He was also anxious to develop the facilities for the teaching of geography, especially to be able to cater appropriately to boys who would in later life find employment in different parts of the Empire, or in other countries.[16] He reorganized the school curriculum, to bring it as he perceived "into closer touch with the change in the educational system which has been going on during the war and since," but was careful in doing so to protect the established position of the classics.[17]

Examinations were also emphasized, and the numbers of scholarships gained to Oxford and Cambridge soon became as impressive during his time at the school as they had been at Bristol Grammar under his headship. In 1925, 30 scholarships were gained by pupils at Marlborough to Oxford or Cambridge. However, there was a basic tension within the school between athletics, which were cherished, and academic and literary pursuits. This had been a characteristic source of conflicts within the public schools since the second half of the nineteenth century, but the tension was present at Marlborough in an especially virulent form. Norwood found himself needing to mediate between two rival traditions, athleticism on the one hand and aestheticism on the other.

These disputes were fluently recalled in the memoirs of one former pupil at Marlborough, T.C. Worsley.[18] The fourth of five children in a parson's family, Worsley had won two scholarships to Marlborough, the lower one open to all sons of clergymen. He was successful in sports at Marlborough and was in the Cricket Eleven for three years as well as in the Rugby Fifteen in his final year. He was therefore in a privileged position to witness, though not—as he admitted in retrospect—to *challenge*, the social antagonisms over which Norwood presided. At the heart of these conflicts, on Worsley's testimony, were the frugal conditions of life at Marlborough. "Marlborough prided itself on its toughness. The amenities of life were non-existent: life was lived on the barest of bare boards, at the smallest and hardest of desks, in the coldest of cold classrooms, in a total absence of any possible privacy. One was always cold, usually hungry."[19] Sports thrived in this climate, and yet academic aspirations also managed to survive. Worsley explained the continuing role of intellectual pursuits as being due to the social composition of the school: "Being a foundation largely for the sons of clergymen, a high proportion of the boys came from bookish homes; and there was always a small hard core of intellectuals at the top of the school on its intellectual side."[20] Nevertheless, Worsley continued, the athletes were far stronger than the aesthetes both among the boys and the masters, and it was only Norwood's intervention that afforded the more scholarly and intellectual pupils an element of protection.[21]

In many ways, these hostile conditions encouraged the emergence of an intellectual counterculture. This was represented at Marlborough in three pupils who were to make notable marks in their later lives. The first was John Betjeman, later to become Poet Laureate, who went to Marlborough in September 1917, aged 14. Betjeman always recalled his schooldays at Marlborough with distaste, and indeed his chapter on Marlborough in his autobiographical work *Summoned by Bells* evoked a gloomy and threatening atmosphere of "Doom! Shivering Doom!":

The schoolboy sense of an impending doom
Which goes with rows of desks and clanging bells.
It filters down from God, to Master's Lodge,
Through housemasters and prefects to the fags.[22]

Betjeman stressed the strength of the chapel in Marlborough's corporate life, but more sinister and potent to him was the "greatest dread of all, the dread of games!"[23]

As he recalled it,

> Great were the ranks and privileges there:
> Four captains ruled, selected for their brawn
> And skill at games; and how we reverenced them!
> Twelve friends they chose as brawny as themselves.

These sat "lording it" in "huge armchairs beside the warming flames," while "The rest of us would sit Crowded on benches round another grate." Moreover, while Betjeman admired the athletes, he could never join their ranks—for he had the misfortune to have intellectual tastes:

> Upper School captains had the power to beat:
> Maximum six strokes, usually three.
> My frequent crime was far too many books,
> So that my desk lid would not shut at all:
> "Come to Big Fire, then, Betjeman, after prep."[24]

Betjeman also had no interest in rugby, cricket, or running, and so he was a natural member of the downtrodden yet spirited group of aesthetes at the school.[25] Other pupils were not impressed by the young Betjeman's attitude. One later recalled that "Betjeman was a soft suburban type who did not like Marlborough, while the rest of us never wavered in our partiality for it as the best of schools in the most civilised part of the globe."[26]

A fellow rebel was Louis MacNeice, who was also to go on to become one of the major English poets of the twentieth century. The son of a rector, MacNeice entered Marlborough in September 1921 and found it difficult to adjust to the hardships and discomforts that he encountered there.[27] In his memoirs, *The Strings are False*, MacNeice showed more forgiveness than Betjeman to the environment that had nurtured him, but he too recalled the bitterness of the feuds between the aesthetes and the athletes. He himself was "ill-qualified for social-climbing and not an athlete."[28] He was fascinated in the arcane rites of Upper School that reflected the prevalence of a cruel authoritarian culture of "government of the mob, by the mob, and for the mob."[29] Nevertheless, he noted, unlike many public schools, Marlborough had what he described as a "strong highbrow tradition," through which "there was always a group among the older boys that was openly against the government, that mocked the sacred code and opposed to it an aesthetic dilettantism."[30]

A third leading member of this notable group of dissenters was Anthony Blunt, who entered Marlborough in January 1921. Blunt was to become a prominent art historian and was appointed as surveyor of the King's Pictures and also director of the Courtauld Institute of Art in London. In 1979, he achieved notoriety when he was exposed as having spied for the Soviet Union during World War II.[31] Blunt's family and social background was in the Church, with his father an Anglican vicar. He loathed sports, and at Marlborough, according to his biographer Miranda Carter, was "useless at games—a prerequisite, for younger boys at least, for school success—treated as a freak, viewed as a loner."[32] As he progressed he found common cause with other self-styled *enfants terribles* such as Betjeman and MacNeice. Blunt was already on the account of MacNeice's memoirs the "dominant intellectual" of the group, who "preferred Things to People" and cultivated "a precocious knowledge of art and an habitual contempt for conservative authorities."[33]

It fell to Norwood as master of Marlborough to attempt to reconcile the rival cults of athleticism and aestheticism. He was clearly uneasy at the excesses of the dominant athletic tradition. He opposed its emphasis on the inculcation of individual competition, in favor of an ideal of service to the community. Indeed, according to Norwood, "team games are played in order that you may learn to serve your side, to combine, and to avoid selfishness: in proportion as games lead to purely individual glorification they cease to be of value to a school which is holding its true traditions."[34] Thus, he continued, "when a school sets itself to producing athletic champions, it is mistaking the means for the end, just as badly as the Germans did in their worship of military force, and just as badly as modern industrialism does when it sets up money."[35] Looking back some years later on the nature of education at Marlborough, Norwood acknowledged that in this "self-contained community," it was "always possible that athleticism would become too dominant, for the tendency of youth is ever to deify the athlete." This was especially the case in some of the Houses, as he recalled, although "Hellenism always seemed to me capable of keeping its end up against the Philistine."[36] He was much too complacent in concluding that these opposing tendencies had been successfully reconciled: "Setting aside the individual cases of boys unfitted by their own nature for community life, I should not say that there was ever a time when those of special gifts and interests would have found themselves thwarted, and their lives made unhappy."[37] Nevertheless, his instinct as master was to try to protect the aesthetes and intellectuals, and to be tolerant of their defiant protests.

Tolerance was therefore Norwood's initial response when Blunt established an alternative school magazine, characteristically entitled *The Heretick*, in March 1924. The heading of the first issue declared boldly: "Upon Philistia will I triumph," although its editorial echoed Norwood's approach in seeking a compromise between athleticism and aesthetic intellectualism, and in avoiding excessive individuality.[38] In criticizing both athleticism and intellectualism, however, it ran the risk of offending both parties, as was certainly the case with one article in this first issue, entitled "The Road." In this article, the narrator is accosted on his walk on a road through unknown territory by two strangers. The first is an old man who invites him to read a list that he has compiled, recording every occasion on which the word "scilicet" occurs in Latin literature. Rebuffed, the stranger then sinks into quicksand. Then the narrator comes across a middle-aged man "dressed in a fantastic garb and evidently addicted to childish pursuits although having reached an age when he should have known better." This newcomer proposes to teach the narrator how to play Rugby football and claims that "This game while providing a maximum of unpleasantness in a minimum of time, and giving free play to all the brutal instincts of human nature, yet so ennobles you that it is almost the only thing you can do to be saved."[39]

Not surprisingly, some parents of pupils at the school began to register strong concerns at such irreverence, but Norwood continued to reserve judgment on it. He was observed carrying it under his arm while watching a cricket match, and, according to one account, began a sermon by saying, "A magazine has appeared in our midst bearing upon its cover the superscription 'Upon Philistia will I triumph.' If this means that we will triumph upon the Philistine within ourselves, it is well. But if it is only another piece of intellectual snobbery, it is not well."[40] However, the second issue of the magazine went too far, for example, by questioning the idea that art could be either moral or immoral, and by advancing the principle of socialism as an ideal of service designed to break away from the "foolish and inefficient" system of industrial capitalism. It even went so far as to make a disrespectful allusion to a deity known as the "Great Sun-God Cyril."[41] The magazine was closed down without further ceremony.

"The Boot"

The controversy over *The Heretick* effectively highlighted the ambiguities of Norwood's own position and the limitations of his authority at Marlborough College. To a large extent, he did retain an impressive presence during this period. He was conspicuously successful in developing its academic profile and maintaining the strong track record that he had established at Bristol Grammar School, and he won widespread respect among the pupils for this. One former pupil, G.V.S. Bucknall, later remembered Norwood as "an impressive figure," with a "kindly warmth" under his "seeming impassivity." Bucknall gave Norwood credit for improving the academic standards of the school, and replacing St. Paul's School in London at the top of the awards ladder. In both 1925 and 1926, according to Bucknall, "I had the unhappy distinction of being the only member of the Classical Upper VI not to win a scholarship or exhibition at Oxford or Cambridge." After his first failure, Norwood apparently persuaded Bucknall's parents to allow him to stay on for another year to try again, but "again he flew me too high."[42] According to another former pupil, too, "He obviously valued the results highly, because I think in my last year, 1923–4, we gained a record number of Oxbridge awards—24 I think—and School Certificate awards were taken for granted. I know when I got an award to Cambridge, I had a charming note of congratulation on the award, 'if somewhat unexpected!' "[43]

Norwood also had an imperious bearing that made a strong impression on those who came across him. He may have looked, in the words of one witness at this time, "like a policeman in an early Chaplin film,"[44] but he made the most of his physical qualities in addressing his pupils. According to John Bowle, a contemporary of John Betjeman, "Norwood was in fact, under a terrifying exterior, a man of, I'd say, something like genius. He certainly had a very great impact on all of us, including John. He had a wonderful voice and his sermons I can still remember . . . . He was a formidable man out for efficiency, with a certain moral purpose without cant in it." At the same time, Bowle added, "Underneath his forbidding exterior, Norwood concealed a great sense of humour and great perceptiveness about character."[45] Similar comments were made by other former pupils at Marlborough, such as one who became a schoolmaster at another independent school for many years and described Norwood as "a lifelong role-model of mine."[46]

Norwood also continued to present lectures for the whole school, as he had in Leeds and Bristol, and these were generally celebrated and memorable occasions. In

February and March 1924, for example, he delivered two lectures on the authorship of Shakespeare, which he had previously expounded upon at Bristol Grammar School. In the first lecture, he presented the argument that Shakespeare was not the author of the plays usually associated with his name.[47] In the second, he set out to refute the case set out in the first, finally confessing that he himself found it very difficult to reconcile Shakespeare's known character with that of the author of his plays, but that he found the evidence in favor of Shakespeare's authorship so overwhelming that he was forced to believe it.[48]

On the other hand, as Norwood advanced into middle age, he became much less active in the life of the school and tended to retreat into the privacy of his study. Reasonably enough, he became a spectator rather than an enthusiastic participant at sports events. He also took less interest in debating and other school societies, and while he still taught senior pupils in the classics there were many who never came across him at all. According to one former pupil, John Baines, "I myself had never come into contact with him but legend credited him with knowing the names and faces of most Marlburians, and some evil-doers hoping to escape notice by sheer anonymity had been taken considerably aback on being addressed by their correct names."[49] Norwood's increasing remoteness probably added to his mystique, but he seems to have been an aloof and rather mysterious figure for most pupils at Marlborough. One former pupil later recalled him as "an aloof Olympian figure, presiding at formal occasions."[50] Another noted that Norwood "was a remote figure for a little boy and I can't recall ever having conversations with him."[51] He was, according to testimony from a further Old Marlburian, "quite out of our reach," uncommunicative and distant: "On reflection I cannot have found much to distinguish the Master of Marlborough from the Old Testament God whom well brought up children were taught to fear and occasionally to worship."[52] Thus Norwood underwent a key shift in his outward persona during this period, linked with his elevation to the heights of Marlborough College, from the ubiquitous and even gregarious colleague that he had been in his earlier career to the reticent and inaccessible figure that he was to represent thenceforward.

Nonetheless, Norwood's social background and grammar school experience led him to be regarded socially with some disdain by many of the pupils of Marlborough. There was a social gulf between the public school and the grammar school that Norwood's presence did little to bridge. According to MacNeice's autobiography, schoolboy snobbery was endemic during this period: "Their conversation, when it was not shop, was infected with the social snobbery they brought from home." MacNeice recalls that such attitudes extended to judgments about the clothes worn by pupils:

> During term we wore uniform black but at the end of term we were allowed to wear ordinary suits to go home in. At the end of term accordingly everyone was jealously competitive and those boys were despised whose clothes were not well cut. As for the boys who went home in their school clothes—of whom I was one, for I was ashamed to ask my family for a decent suit—they were almost pariahs.[53]

Masters were certainly not exempt from this kind of snobbery, and many pupils viewed Norwood himself as socially inferior.

It seems to have been this sense that Norwood was "not quite-quite" that led to him being given the nickname of "Boots" or "The Boot" during his time at Marlborough. As T.C. Worsley recorded in his autobiography, "He was called 'Boots,' not so much because he habitually wore them, as because they were the outward and visible sign of his not being quite-quite." Indeed, Worsley continued, "He had been appointed to Marlborough from some minor grammar school, and such was the snobbery of us well-brought-up children that, just for that reason, he was booed in the school hall on his first appearance."[54] One former pupil also suggested that Norwood's wife, the ever loyal Catherine, was known as "The Buttress."[55] Some pupils did not associate Norwood's nickname with a social stigma. Christopher Bell, for example, assumed that he was known as "The Boot" "perhaps because he still favoured this type of footwear at a time when fashion was strongly towards the shoe."[56] Yet there was a definite overtone of social class difference in the use of this nickname. Another pupil, M.J. Hayward, an admirer of Norwood's educational methods, could not refrain from mentioning in his unpublished memoir that Norwood had arrived from the "foreign milieu" of Bristol Grammar School and had "once been seen wearing button boots."[57]

Thus, there were a number of tensions current at Marlborough College that Norwood was unable to resolve. While he emphasized the traditions of the school, the clash between aestheticism and athleticism was troubling and discordant. The academic successes that he also encouraged did little to reduce social snobbery, or the social distance between public schools on the one hand and grammar schools on the other. Although he remained an impressive and formidable figure, too, he became perceptibly more distant and remote at this time. On the other hand, he could point to a generally strong record in his management of Marlborough, and he appeared to have settled there for the foreseeable future. His relations with the staff, pupils, and the governors were fairly harmonious. His family life became connected with Marlborough in a tangible way in 1921 when his eldest daughter Enid married one of the housemasters, Clifford Canning, who was considerably older than her and had been a master at Marlborough since 1906.[58] An inspection of the college by the Board of Education in May 1924 was categorical in its finding that "The Master is thoroughly fitted for his important post by scholarship, teaching power, personality, and wide educational experience. His administration and control of the School have been singularly successful."[59] Therefore it was unexpected when at the end of 1925, Norwood announced that he was leaving Marlborough to take up the headship of Harrow School. He did so under pressure to improve the declining fortunes of an ancient and prominent public school. He soon found that the tensions he had encountered at Marlborough were very minor compared with what he was to face at Harrow.

An Alien at Harrow

When Cyril Norwood was appointed in December 1925 as the new headmaster of Harrow School he was, at the age of 50, at the pinnacle of his highly distinguished educational career. His appointment brought with it renewed recognition and enhanced status. His photograph was featured on the cover of the magazine *Country*

Life, which wished him "all good fortune and success" in his new appointment. It noted that he was unlike previous heads of Harrow, in that he was not in holy orders and his experience was not confined to public schools and universities. If these qualities might give rise to anxiety in some quarters, it added that Norwood possessed "a knowledge of men and affairs uncommon among schoolmasters." Moreover, it pointed out, "he has shown himself always a student without rigid and fixed ideas, ready to learn by teaching."[60] These high hopes were to be largely disappointed. After eight difficult years at Harrow, Norwood resigned his post to become president of his old Oxford college, St. John's. He remained there until his retirement, in 1946, and never held another headship. The problems that he encountered at Harrow were partly personal in nature but were also social and ideological as they involved an explicit contest between rival and opposed conceptions of the public school tradition.

After Norwood's death in March 1956, some lingering echoes of controversy about his time at Harrow could be discerned through the obituary columns of *The Times*. Norwood's obituary referred to the "disillusionment" that he had experienced in his years at Harrow,[61] and in the days that followed, other sympathizers added their own personal testimonies. One of these noted that a "struggle" had taken place between Norwood's ideas and "traditionalism."[62] The obituary that appeared in Harrow's own magazine, *The Harrovian*, made a point of mentioning what it described as the "not very creditable obstructiveness" that some of his staff had shown to his reforms at the school.[63]

More hostile voices, while they also recognized the difficulties of the time, tended to blame Norwood himself for exacerbating them. One former pupil, Giles Playfair, complained in his autobiography (published in 1937, when he was 26 years old) that Norwood was "an excessively self-opinionated man" whose attitude was "tyrannic," and that he "never welcomed contradiction or allowed his will to be flouted." Moreover, Playfair continued, "Dr Norwood did not live up to expectations. I have a suspicion that his real trouble was that he never got used to Harrow. He never really liked it. His heart remained in Marlborough."[64] A recent major history of Harrow School, produced by Christopher Tyerman, reaches a similar verdict. This book provides a candid appraisal of the conflicts that took place under Norwood, and it concludes uncharitably: "Typical of self-confident and not unsuccessful self-publicists, Norwood was honoured less by those working closest to him than the outside world . . . Privately affectionate and concerned, he was viewed by snobbish boys and masters as common, dictatorial, insensitive, a poor listener, and in his views on education, fairly absurd . . . ."[65] Tyerman thus portrays Norwood's headmastership as an assault on Harrow's distinctive traditions.

While there were clearly both supporters and detractors to be found, Norwood himself was clearly alienated by his experience at Harrow. When the opportunity to leave for St. John's came to him, he sat down and wrote two lists of reasons with his wife, Catherine, one for staying at Harrow and the other for leaving it. In favor of staying were financial security, the familiarity of the work, his dislike of the climate in Oxford, his suspicion that he would be confined to official business rather than be involved with students, and the prospect that he would be perceived as having been defeated by his enemies at Harrow. On the other hand, he noted, St. John's would probably provide "for the next 12 years, and probably 17, if health and strength are given, a position of comfort and dignity with abundant leisure." Moreover, Catherine hated Harrow and was, on her own account, treated like a "drone," and regarded "another seven years with dread." Indeed, he added, "I too am an 'alien' at Harrow. It has never been a part of me as Bristol and Marlborough were." He clung to the view that if he left, "Harrow would regard it as a disaster, and I have reason to be grateful to many." But his wife had a decisive answer to this: "To whom have you to be grateful? Your staff? Friends—you have none."[66]

While Norwood's difficulties at Harrow were partly due to the strength of the resistance that he encountered to his reforms, they also owed much to social class differences and in particular to the elitism characteristic of many of the staff and pupils. He found it very difficult to gain acceptance at the school from some of the boys and masters, and this was at least in part because of the high social status of the school that he could not himself claim. There is an interesting glimpse of this in Jonathan Gathorne-Hardy's book *The Public School Phenomenon, 597–1977*. Gathorne-Hardy suggests that Norwood's earlier appointment as the head of a grammar school led him to be despised by "snobbish pupils" and "snobbish staff," and he quotes an unnamed informant from Harrow in the 1920s who says that Norwood was brought in to "clean Harrow up": " 'But he failed,' said my aged informant with considerable venom, 'because he wasn't really a gentleman. His nickname here was "Boots"—not quite-quite. He never got Harrow into his grip. He retreated to write his ridiculous book, to sit on Whitehall committees and eventually to St John's College, where he made an even worse mess.' "[67] The stigma that had first attached itself to Norwood at Marlborough had followed him to Harrow, if anything in a more virulent form.

When Norwood was appointed as head of Harrow, he was under no illusions as to the magnitude of the task that faced him. Harrow retained an imposing reputation and could claim the prime minister (Stanley Baldwin), the chancellor of the Exchequer (Winston Churchill), and the archbishop of Canterbury (Randall Davidson) among its former pupils, but influential observers feared that his reputation was in decline. The previous head, Lionel Ford, was generally regarded as having been too weak to cope with factions among the staff and the behavior of some pupils. Matters came to a head after some of the school's pupils, including the headmaster's son and head monitor, were discovered enjoying themselves in a London night club. Cyril Alington, the provost of Eton College, another public school of the front rank, helped to persuade Norwood that he should accept the challenge of Harrow by insisting privately that he represented "the only chance for Harrow." Like Norwood, Alington was the son of a clergyman. He had himself been educated at Marlborough before teaching there and had gone on to Eton, where he became the first non-Etonian headmaster of Eton for over two hundred years.[68] He therefore had some understanding of the kinds of pressures that Norwood would face, and in addition to this, he was aware of the difficult period that Harrow School was experiencing. Alington continued,

> I know a good deal of the local situation and I confess that I think rather badly of it: they have shown a persistent inability to appreciate L.F. [Lionel Ford] and a very curious ingratitude for all that he has done for them. There is a danger of reaction against religion on the ground that he has pushed it too hard. If they get a weak inexperienced man now they will collapse into the slough of despond from which he with difficulty has been pulling them, and I know no man except you who could tackle the job with a real prospect of success.

He went on to warn that it would be a most difficult undertaking: "It would no doubt be very difficult to start it on a new line, but I think you could do it after you had wrung the neck of some Old Harrovian Masters and a great many Old Harrovians outside." Therefore, Alington concluded rather melodramatically, "I have no doubt from any point of view, professional, personal or pious, that you've got to go and save Harrow!"[69] It was in this spirit that Norwood accepted the post.

Harrow School was established on the basis of a charter obtained by John Lyon from Queen Elizabeth I in 1572, and developed as a Free Grammar School after 1615. It rose to national eminence in the eighteenth century and attained new levels of success under Charles Vaughan (1845–1859).[70] In the 1920s, Harrow had over six hundred pupils at any one time, all boys, each of whom paid an annual fee of £210 with additional costs of £50 to £75 per year. Boys were admitted to the school at the age of 13 and usually remained until they were 18 or over.[71] About half of the pupils went on to higher education, and most of them (i.e., 40 percent of all of its school leavers) went to Oxford or Cambridge.[72] Many of the pupils were from influential families and would themselves go on to take a wide range of leading positions in politics and society. These included the future financier Victor Rothschild (entered Harrow in 1924), the future playwright Terence Rattigan (1925), Tom Harrisson (1925) who went on to found the influential organization Mass Observation, and the future King of Iraq (1926).

Norwood set about reforming the curriculum of the school in his characteristic style. He retained the same subjects that had been taught previously, but a Board of Education inspection of Harrow held in October 1931 found that the changes he had made had been "important and far reaching." In particular, the position of the classics was enhanced, with Norwood himself taking part in teaching the classical sixth form. On the other hand, provision for science was reduced so that while more pupils learned Latin, fewer learned science than before Norwood had been appointed as headmaster. The Board inspectors noted that a large number of boys omitted science from their choice of subjects, and felt obliged to suggest that this omission constituted a loss, "not only or chiefly, to the boys who will later join the Science Upper Fifth, but also to those who will pursue higher studies of other kinds."[73] This problem was exemplified in the experience of a Harrow pupil, Anthony Part, who was to become a leading civil servant. Part recalled in his autobiography that he studied virtually no science either at his preparatory school or at Harrow, and that mathematics also ended at School Certificate stage. He complained in retrospect that this lack of balance between the arts and the sciences was "a nonsense" in terms of preparing for an increasingly technological world:

> My own career turned out to involve spending many years working alongside various kinds of engineer and scientist as well as architects and quantity surveyors. A good supply of wet towels was needed to master enough of the technology to become an effective partner. Learning some of the elements at school would have been much more satisfactory.[74]

There was little prospect of such a preparation at Harrow under Norwood.

It was evident that Norwood himself was not sympathetic to the cause of science. This was reflected, for example, in the position of A.V. Hill, who at the end of 1928 was appointed by the Royal Society as its representative on Harrow's governing body.

The chairman of the governing body, Francis Pember, welcomed him particularly for the encouragement that he would provide for the science masters at Harrow.[75] Yet, by the summer of 1930, Hill had already seen enough and wanted to step down from this position. He was persuaded to stay, but toward the end of 1933 he again requested that an alternative be found.[76] He explained his reasons for wanting to resign in terms of his frustration with Norwood's approach to the science curriculum. In general, he noted, "my own lack of experience in legal and business matters makes me of no use at formal meetings of the Governors, while I have found little other means of being of service." He had been unable to discuss the further development of the science curriculum with Norwood: "The Head Master has never consulted me on matters connected with the teaching of science or mathematics in the School, and although I might have spent much more time than I did with other Masters, who always seemed pleased to see me, I doubt if this would have been much use without the Head Master's interest and co-operation." Moreover, he elaborated,

> What was wanted, and what I would have been glad to join in, was a general discussion of the place of the subjects in the scheme of education at Harrow; that place is not I think satisfactory, particularly in view of two things: a) and, perhaps less important, the comparatively small number of boys from Harrow who attain distinction in science or mathematics; b) the fact that it is possible for able boys to pass out into the world of affairs without any contact at all with science. This was not always so.

It was a pity, he averred, "that the abler boys should grow up without any appreciation or understanding of science, when they are expected later to be leaders in a civilisation which is based so largely upon it." Nevertheless, he acknowledged, a headmaster was "Captain of his own Ship," and "the school curriculum is his own affair unless he wishes to make it otherwise." Hill's letter ended with a postscript that it had been written before he had learned that Norwood had resigned: "Perhaps there will be a change of policy and attitude with another Head Master."[77]

From this evidence, it appears that Norwood's preferences and predilections in the curriculum domain were becoming increasingly pronounced, and more rigid, as he became older. This may also have been true for his approach to the management of the school and its staff as he became more distant and remote. Nevertheless, he continued to command respect from the Board of Education's inspectors. The inspectors' report of October 1931 gave fulsome praise to what it described as "the substantial progress of the School, both on the material and on the educational side, since it was last inspected and notably in the last five years." They gave most of the credit for these developments to Norwood himself: "If the School owes much to the increasing generosity of the Old Harrovians and to the work of the assistant staff, it also owes more than can be explicitly stated in this Report to the Head Master, whose services the Governors were so fortunate to secure in 1926 and whose appointment has been the most notable event in its recent history." Moreover, the report concluded,

> It would be difficult indeed to point to any important department of the life and activities of the School which has not felt the impress of his forceful personality, and the Inspectors to whom his work in former posts was already known, would be failing in

their duty if they did not bear witness in this Report to the strength and sanity and judgment that the Head Master has brought to the conduct of the School's affairs.

Finally, at a meeting between the inspectors and the school's governing body, in the absence of the headmaster, the inspector F.B. Stead embellished this praise further: "he had known the Headmaster for a long time; during the inspection of Harrow he had been impressed more than ever by the Headmaster's strength, sanity and judgment."[78]

In most cases, too, the pupils at the school were impressed with the qualities they found in their headmaster. Some had the opportunity to see him at close quarters, especially those whom he taught in the sixth form or those whom he confirmed for the Church. One Old Harrovian who was taught classics by Norwood at this time remembers him as "a 'hands-on' headmaster," who taught the Classical sixth for an hour every day (two hours on Thursdays), including three periods at 7.30 am: "Even at the age pf 17/18 it struck me as remarkable that in spite of all the 'admin' that must have gone with his job, his scholarship was still unimpaired and meticulous."[79] Another, who was confirmed in the Church by Norwood, recalls that

> In fact Norwood was a deeply Christian man and I do have a very lasting memory of his preaching to those who were to be confirmed—it was the last part of our preparation for that ceremony; I don't remember what he said but I have always remembered the effect his message had on me; I was profoundly touched by his call to a truly good—say holy—life style—in morality, ethics and general obedience to God.[80]

For others who came across him only at ceremonies or in the chapel, he had a "presence," as one explains:

> Occasionally he delivered a sermon. It was like the wrath of God! Aged about 14 I had a room mate with whom I shared a Kelley's Keys to the Classics which were forbidden but without which Latin translation took a lot of one's time. One summer Sunday he gave a sermon on the evils of cheating. When I got back to the House I found my roommate had been so impressed that he had burnt our cribs. They cost a lot to replace![81]

In these ways Norwood continued to exert a strong influence on the life and development of the school.

Nevertheless, it is the unapproachable and distant nature of Norwood as the head of Harrow that is striking in the testimony of many Old Harrovians of this period. One remembers that he was "the personification of authority," with both masters and boys "in awe of him," and that he exercised a "benevolent despotism," but that "it was perfectly possible to spend one's entire time at Harrow without ever being actually in contact with the Headmaster."[82] Another notes, "Norwood was a splendidly aloof headmaster. Very dignified and rather austere. It was difficult to imagine him in any position other than standing or sitting bolt upright."[83] Another recalls him as having a "very imposing presence, in a flowing gown, at functions and very awe inspiring to junior boys." Apparently the wives of the masters were "terrified" of him, to the extent that if he was on the platform of the train station going to London and they saw him, "they fled to the back of the train to avoid travelling with him."[84]

Others emphasize the remoteness of Norwood's persona and his lack of personal contact:

> I never met him. I imagined that this was quite normal at large Public Schools. In fact the only time I saw him (as far as I can remember) was when he preached to us in chapel, which was not very often. . . . But I would say he was a bit of a recluse.[85]
>
> . . . he remained a remote and respected man with whom one had no contact of a personal nature even as a School Monitor.[86]
>
> I never came in contact with Dr Norwood in any way—my recollection of him therefore is as someone rather distant, who one knew by sight, but not much more.[87]

Personal glimpses of the man himself were few and far between. One pupil was privileged to see him "walking down the street with his magnificent dog, that rare breed, a Kneeshond, to whom he was obviously devoted."[88] Another recalled some "restless spirits" in the classical Sixth when Norwood was teaching, he "found Norwood's Olympian manner a little irksome," and began circulating private jokes:

> Norwood was a heavy smoker, and one senior boy (later to become governor of a crown colony in the West Indies) whispered to his neighbours to watch a fly crawling across the wide mahogany table we all sat around. Sure enough, once it came to pass the Head master, it immediately fell dead. Each of us, sitting with him to have our Latin verses corrected, had been aware of the powerful tobacco fumes.[89]

None appears to have had any inkling of Norwood's family background, still less of his father's problems. One had a memory, passed on from one of his own family members, that Norwood considered it to be his duty to be severe toward youthful offenders because of experiences that he had with his own father, possibly something to do with his father's drinking problem.[90] No rumors of this kind appear to have circulated at the time as Norwood kept his distance and covered the traces of his social and family origins. Any such stories would have been highly embarrassing even if he had been at ease and accepted at Harrow. In fact, he was neither, and he found himself pitched into an enervating conflict that threatened his basic educational beliefs.

A Tradition of Freedom?

In the nineteenth century, especially under Montagu Butler, a strong house system had developed at Harrow, a system that consolidated a federal or balkanized structure for the entire school, which was dispersed geographically over a wide area.[91] This led in turn to housemasters becoming very powerful, even though technically they possessed very little security of tenure, while headmasters were constrained in their ability to exercise their authority over the school as a whole. There were about fifty masters at Harrow during this period who were responsible for teaching the school subjects. Among these were the housemasters who had become increasingly influential under the unfortunate Ford. It was these who represented the center of resistance to Norwood's attempts to provide strong central leadership. Norwood associated the character of the public schools with the authority that Thomas Arnold had shown as headmaster at Rugby School in the early nineteenth century. Thus was set in train a clash between Norwood's idealized view of the English public schools, and the particular set of arrangements that had been established and were in some quarters venerated at Harrow itself.

One housemaster who was especially fervent in his support for the established traditions of the school was Charles George Pope. A classics master at the school since 1900, and housemaster of the Grove since 1915, Pope was an old boy of Harrow, captain of its Cricket XI in 1891, and had taken a First in the Classical Tripos at Trinity College Cambridge.[92] He was widely known as "Cocky," apparently a reference to the vigor with which he tried to make the Grove "Cock House" in inter-House sports competitions.[93] Nevertheless, it was his strict adherence to school and House traditions that attracted most comment. An appreciation of his life in *The Times* would later describe him, affectionately, as a "tory of tories."[94] He refused to compromise on his commitment to the customs of his House. Surviving former pupils of the school, looking back over seventy years, vividly recall Pope as an "absolute traditionalist," who "clung obstinately to old ways . . . whether in school dress or the boys' comfort."[95] These cherished habits were evident in the everyday practices of Pope's House. Although enthusiastic about school sports, for example, Pope had very little interest in the Officers' Training Corps.[96] Boys in the Grove wore tailcoats every day, while the rest of the school wore blazers, and they were not allowed to have the basket chairs, or "frowsts," that were general elsewhere.[97]

When Norwood arrived at Harrow in 1926, Pope was trenchant in his opposition to reforms. He regarded Norwood as a menace to the traditions of his House and of the school as a whole and took every opportunity to try to thwart him. He made these feelings known to the boys. One of his former pupils remembers him referring to the headmaster as the "Yellow Peril"—"an expression in vogue at the time, referring to the millions of Chinese who might one day take us over."[98] Another former member of the Grove confirms that Pope "did not get on with Norwood at all; and made his feelings known to us boys (looking back, this seems rather disgraceful)." This old boy adds that Pope, "at that time a master of long standing, and much respected, must have been a thorn in Norwood's side."[99] Over the next few years, Norwood and Pope engaged in a number of battles relating to reform, often over relatively minor matters that assumed symbolic significance for what they appeared to represent. One of the most characteristic of these was over the unique form of football that was played at Harrow. This was "designed to be played in a bog,"[100] and was well suited to the clay soil of Harrow, but it did not allow interschool competition, and so Norwood proposed to replace it with rugby. The philathletic club at the school opposed this, and so Norwood called the entire school for a referendum that he eventually won. The football fields were drained, rugby was introduced for the autumn term, and Harrow Football was moved to the spring term. Such changes were anathema to Pope as they symbolized the undermining of the traditions of the school.

An open feud developed between Norwood and Pope that continued for three years, well into 1929, when Norwood at last found an opportunity to dismiss his fiercest tormentor from the school. The headmaster was deeply offended by an incident in which Pope laid a bet on a horse race (the Derby) on behalf of a member of his House, Victor Rothschild, who was a school monitor. This appeared to Norwood to be an ill-disciplined breach of a school rule, and this incident was followed by another when Pope allowed members of his House to stay with him in Winchester, without permission, during a cricket match against Winchester College.[101] However, it was clear that Norwood had more general reasons for choosing to confront Pope on this issue. He accused Pope of being disloyal and of fomenting opposition. Indeed, Norwood complained, "since I have been Headmaster, you have claimed a position from which you are to be free to criticise, to foster opposition, and to work on the minds of boys so as in effect to make them disloyal."[102] He continued,

> There are certain sides of the school's activities which you foster, but there are others which you bring into contempt and thereby render the work of your colleagues difficult. More than one of the parents in your House have complained to me that their boys are being "taught disloyalty," and have asked me what can be done about it. I have tried to excuse you though I know that they have gone away, thinking me to be weak and afraid.

He was now at the end of his patience: "I can suffer no longer the continuance of the 'imperium in imperio,' nor can I tolerate longer the separatism and contemptuous independence, which is the subject of comment among parents, boys and masters."[103]

Norwood took legal advice from the school solicitor, that masters held their posts under the school's statutes "at the pleasure of the Headmaster." A similar case at Richmond Grammar School in 1907, which had gone to court, had established that the head had the sole power of appointment and could dismiss any assistant masters without leave to appeal. According to the school solicitor,

> The Endowed Schools Acts do not apply to Harrow, but judging from the case I have referred to it seems to me that in strictness Mr Pope is not entitled to any notice, and that the letter you have already written to him complied with the requirements which would have been necessary if the Endowed Schools Acts did apply, and these requirements it seems to me are the only ones that could be brought.[104]

The chairman of the board of governors, Francis Pember, was nervous of the likely consequences and advised caution. Nevertheless, Norwood told the housemaster either to resign or to accept dismissal. When Pope refused to resign, Norwood dismissed him.[105]

Norwood's indignant reference to "imperium in imperio" reflected not only his resentment at Pope's criticisms and disloyalty, but also his awareness of two rival traditions in conflict with each other. Pope himself claimed that "imperium in imperio," far from being a fault as Norwood argued, was in fact "more or less the ideal for a House in a Public School," although he added that he would never assert "imperium contra imperium."[106] For his part, Norwood was insistent that

> There is between us a conflict of traditions which are not to be reconciled, and so long as that which you call the tradition of freedom persists, so long is the H.M. without real authority—and after a certain period he is forced to forfeit respect as well. I do not say that you invented the system by any means, but I think you have whole heartedly maintained it.[107]

Pope took his case directly to the chairman of the board of governors. He professed that he did not understand Norwood's reference to a "conflict of traditions," and could not recall ever talking of a "tradition of freedom." Moreover, he added, "no traditions of my House could possibly impair respect for the Headmaster."[108] He claimed that all he had ever wanted was "to do my best for School and House, and to assist in producing English gentlemen."[109] Norwood, though, would not withdraw, going so far as to threaten to resign as headmaster if the governors did not support his action: "I shall certainly take my stand on the broad ground that your conception of a Housemaster's policy is inconsistent with the position of any Headmaster who is to perform his true function at Harrow, and with the possibility of a united school."[110]

Norwood was as good as his word. His formal memorandum to the board of governors to explain his dismissal of Pope, after recounting Pope's alleged transgressions over the horse race and the cricket match, concentrated on the broader difficulties that he had encountered during his headship—difficulties for which he held Pope chiefly responsible. "When I came to Harrow," he explained, "I found that it was in effect not so much a school as a federation of independent boarding houses, which combined to receive lessons and to play games together." He complained that Pope's House had "advertised its independence and its contempt for school authority by wearing a different dress," while other housemasters "considered themselves free to criticise, to support or not to support, to adopt independent policies." Indeed, "criticism of one another, and of the Head Master, had long been open, and seldom concealed from boys." He found a general opinion that "the old system was entrenched in the Grove and its House Master, that he was too strong to be shaken, and that it was an interesting question to see whether I should 'knuckle under' as my two predecessors had done."

In all the disputes that followed, Norwood continued, "I have had the feeling that Mr Pope has been controlling the strings, and that he has been steadily working through boys on independent and subterranean lines." Indeed, Norwood allowed his accumulated frustrations to reveal themselves as he recounted a long list of disputes large and small. On the introduction of rugby football, he recalled, Pope had opposed the changes at a masters' meeting and had been defeated in a vote by 49 votes to one. Nevertheless, he continued,

> when I came to speak to the Philathletic, I found that they had been "got at," their loyalty appealed to, the Harrow staff derided as men of no importance, the Head Master represented as dangerous and mistaken. I had to appeal from them to the whole school, and even then attempts were made to intimidate the voters. All this, and I believe most of the storm in the papers, was fomented by Mr Pope.

Norwood concluded that Pope could not be trusted and could not change his ways, and that while he remained in his post he would continue to undermine the school itself:

> He is bred in a tradition of disloyalty, and he has been the main support of a disloyalty that has long been a canker at the heart of Harrow, and at one time brought its prestige among English schools very low . . . but Harrow can never be in my opinion what it should be, or what it will be, unless the tradition, for which Mr Pope stands, is openly and completely crushed.[111]

On November 13, the dispute eventually came to a full governors' meeting which Pope was allowed to attend and make a statement for consideration. The school governors decided that "they could not see their way to making any recommendations to the Head Master in respect of the exercise of his authority in this matter."[112] Pope eventually agreed to resign from his post rather than be dismissed, and Norwood insisted on him leaving at the end of the autumn term instead of serving out the whole of the school year.[113]

The unhappy episode had ended with Norwood's authority confirmed and his principal opponent banished from the school. Nevertheless, Norwood was deeply affected by the controversy that had arisen. Some masters, pupils, and parents had either taken Pope's side in the dispute or asked for a compromise to be reached, and it took some time for acrimony to subside. Pope himself continued to protest at Norwood's assault on the Grove and insisted that "The wearing of the black coat was never meant for anything but a lead to the school in the maintenance of the school's own tradition."[114] Although he retired to his residence at Bashley Lodge on the edge of the New Forest (living with a sister who bred prize goats) to "enjoy to the fullest extent his hobby of shooting," Pope continued to take part in Harrow events.[115] Norwood, meanwhile, was worn down by the effort involved in exerting his authority and became increasingly sensitive to perceived slights. In the spring of 1930, for example, he applied to the governors of the school for leave to visit Canada in an expedition of head teachers. Permission was granted, but initially the governors insisted that Norwood should return in time for the start of the summer term. Norwood was deeply upset by this and interpreted it as a criticism of his leadership of the school. Such was the concern of his wife, Catherine, that she brought his disappointment to the attention of the chairman of governors, Pember, who arranged an extension to the Canada visit. Pember reassured Mrs. Norwood that "there is no foundation whatever for the H.M. thinking that he is being estranged here: my admiration for him, and trust in him are as complete as ever." Indeed, he added, "It is perfectly monstrous that any should construe the decision of the Governors as a slap in the face to the H.M." He added that he had himself never disagreed with Norwood on any matter of importance—"except on this one question."[116] What Pember described as Norwood's "stoical courage" led him to carry on his duties, but Norwood never again felt that he enjoyed the "entire confidence and regard" of the governors or the school.[117]

Norwood was thus challenged from within his own school by an alternative and competing notion of the public school tradition. From the perspective of Harrow, Norwood's ideas were forward looking and even dangerously advanced in their willingness to do away with some of the most cherished and distinctive symbols of the institution. Norwood was concerned about adapting to wider educational and social changes, while continuing to insist on the enduring value of the underlying tradition of the public schools. He was convinced that this kind of approach was essential in order to "save Harrow." However, to the diehards such as Pope who were associated with schools such as Harrow, such an outlook amounted to a betrayal of the true tradition, and Norwood became an object of hatred, a symbol of the "Yellow Peril" that threatened to engulf the public schools and everything they stood for.

It is also clear that when he was opposed in his own school, Norwood took his cue from Thomas Arnold, not only in the way that he sought to regenerate the values of the public schools in a rapidly changing social and political context, but also in his emphasis on the authority of the headmaster within the school. Arnold, as Honey pointed out, had invested the role of the headmaster "with a *charisma* which has deeply affected the English notion of the school (and the headmaster) down to the present day."[118] It was at the center of the school community and at the source of its values. Norwood was determined to maintain this key aspect of the public school ideal, even at Harrow, which had developed over the previous century in a different way. It was for this reason that the doctrine of "imperium in imperio," which Pope came to represent was unacceptable and deeply repugnant to him. Individual Houses and masters could not override the authority of the headmaster without destroying the basis of the English tradition itself.

Although he had not managed to "openly and completely crush" the tradition with which Pope was associated, he had done enough to vindicate his own beliefs, at least for those who sympathized with him. In August 1931, he wrote candidly to one of his former pupils at the school: "That you can say that Harrow is a different place consoles me for all the anonymous abuse which I have received from the disgruntled since I have been at Harrow: thanks to the present generation I feel that the fight for the soul of the school—for it was no less—is now a winning fight."[119] When he resigned from his post, he was anxious that he would be thought of as "having really been beaten by Harrow," although his wife insisted that "Harrow knows, and you know that you have not been beaten."[120] The prospect of returning to his old college in Oxford was certainly an attractive one in these circumstances. He was frank when he responded to initial enquiries from the college:

> The position at Harrow (please regard this as very private) is that I am in my eighth year, and have had to break down not only bad traditions among the boys, but also worse and more strongly entrenched traditions among the staff. Here I have had no bed of roses, but at the same time I think this work is done.[121]

At the same time, he added, "I feel now that I have been a head master since 1906: I am 58: I begin to lose the freshness that schoolmasters ought to have. I have at times to force interest in what used to be spontaneous. I shall not feel I am letting Harrow down by going."[122] There was in this a sense of achievement in difficult circumstances, but also of fatigue and closure.

There is also a strong sense of insecurity in his position in the way that Norwood responded to the challenges posed at Harrow. His headship was endangered especially as he interpreted the opposition of Pope in terms of basic principles, and he might well have felt obliged to step down from his post. In a way, he found himself in a position that was as lonely as his father's at Whalley Grammar School fifty years before. The social differences between the public and the grammar schools, and between himself and many of his staff and pupils, probably also helped to heighten this sense of insecurity. Those at the school felt threatened by his reforms; he saw a danger of traditions crumbling if they were not shored up. This was the background to his major work, published in 1929—*The English Tradition of Education*.

Chapter 7

The English Tradition of Education

Cyril Norwood's *The English Tradition of Education* was his most substantial single-authored published contribution to educational debate and was published in 1929 at the height of his influence and reputation. *The Higher Education of Boys in England*, besides being jointly written, had been an early sketch of general issues; the Norwood Report, formally the product of a committee, was a final flourish. It is to *The English Tradition of Education* that we should look for Norwood's fullest and most mature consideration of education. What credentials he has as an educational thinker and philosopher, and indeed as a historian of education, must rest principally upon this work. It deserves attention also as an eloquent expression of the anxieties that surrounded secondary education in the 1920s and 1930s. It defined and celebrated an English tradition as a means of defending established values and structures against contemporary threats. It effectively refused to acknowledge the changing needs or the rights of the majority of individuals, nor the new requirements of a modern industrial and commercial nation state, by insisting that they could be incorporated or subsumed into an existing tradition. In this sense, it was a conservative document and formed part of a defensive project.[1] In addition, it attempted to reorient a number of key debates to respond to current concerns in particular areas. This was again largely to preserve what he saw as most important and precious in the English tradition as a whole, a tradition that, however, included radical viewpoints that put him at odds with entrenched interests.

The personal, educational, and political context of this treatise helps to explain why he wrote it, the character of the work, and the impact that it had. Norwood's public position as the head of one of the most well-known public schools in the country and as the chairman of the Secondary Schools Examinations Council gave him a unique platform from which to expound at length on the problems facing education. Writing such a work gave Norwood a useful opportunity to help to consolidate his position not only as a leading head teacher and educational policy maker, but also as a public figure. Furthermore, it allowed him to respond to the challenges that had arisen to threaten his ideals and values. As we have seen, the expansion of secondary education and the proposals of the Association of Headmistresses and of the Hadow

Report of 1926 led him to express grave forebodings for the future of secondary education. There were also increasing public criticisms of the public schools themselves. Moreover, even within the ranks of the schools in which he was involved there was opposition—to some extent at Marlborough, but especially at Harrow—to the kinds of developments that he favored. *The English Tradition of Education* was intended as a riposte, even as a rebuke, to these varied critics and dissenters.

There is an irresistible parallel to be drawn between Norwood's book and the paean of praise to the Indian Empire that his father had produced in the 1870s. The Revd Samuel Norwood, surveying the chaos of his rural grammar school and the ruins of his hopes and ambitions, had resorted to a celebration of the imperial tradition, partly as a consolation but partly also to provide an inspiring message for the new generation to follow. Cyril Norwood was not in such desperate straits, but like his father, he sought to invoke the past to redeem and regenerate his ideals in the face of his many opponents. Similarly, the idealized and mythologized past that Cyril Norwood produced served to rationalize the structures, values, and relationships that had survived into the present.

Norwood's book also adds further weight to the argument of C. Wright Mills about the importance of connecting private troubles with public issues.[2] Read at face value, it might appear to be a work that identifies him as the heir apparent of the English tradition of education. Contextualizing it in relation to his own personal and family background, his professional difficulties, and the dilemmas that he faced in terms of educational policy, reveals it to be an expression of the antagonisms and struggles in which Norwood himself was an embattled protagonist.

Public Schools and the English Educational System

The growing role of the State in English education during the first three decades of the twentieth century led to a number of attempts to explain the changed layout of the educational system. The position of the public schools in relation to the state system also became the focus of renewed attention by the 1920s. From his new vantage point at Harrow, Norwood made several contributions to this kind of commentary on English education in a wide range of periodicals and magazines. These generally defended the character of the public schools, but they ventured well beyond this to offer a coherent account of the system as a whole.

Among Norwood's interventions of this type was a short article in the *Journal of Education* on the public schools, the fifth in a series of articles by different authors on the schools and universities of Great Britain. This provided a brief sketch of the historical development of the public schools. He observed that these schools included day schools as well as boarding schools, but he concentrated mainly on the boarding schools, suggesting that "the boarding schools are a peculiar product of this country, winning for themselves devoted loyalty and uncompromising hostility, but taking a place in the national life to which no other country offers a parallel."[3] Their origins, as he pointed out, were in medieval and Tudor times, and he argued that their distinctive ideals were already being cultivated from this early period. In particular, he emphasized the contribution of William of Wykeham in founding Winchester College in 1382: "He established the connection between the public school and the university, and shaped the beginnings of the prefect system when he gave to the oldest and most advanced scholars the right to control and teach the younger." He also encouraged a strong corporate ethos and aimed for the development of character in the pupils at the school as was reflected in his famous motto of "Manners makyth man." Norwood also noted the role of John Colet as the founder of St. Paul's School, especially for his humanist values, and concluded that the ideals represented by these early schools and their founders were "the germs of all that has been most vital and most fruitful in English education."[4]

Norwood's historical narrative in this essay skipped over the "depressing period" of the decline of the public schools in the seventeenth and eighteenth centuries, but it dwelled on their revival and expansion in the nineteenth century under the inspiration of Thomas Arnold. He explained the growth of their prestige and influence as being due to their distinctive emphasis on the training of character, although he pointed out that they were helped by a number of social and economic factors including the spread of the British Empire, the growth of national wealth, and the construction of railways. He also acknowledged aspects that tended to undermine the schools. First among these was their sense of social superiority over other forms of education, as he observed, "Largely for social and snobbish, rather than true educational reasons, the boarding-school has come to be thought of as giving an education preferable to that of the day-school. Class distinctions have been created and emphasised by those who have collected money, but have not inherited culture or the tradition of service." Second, he reported that the training of character at times degenerated into a worship of games. Thus, he lamented that "Philistinism and athleticism are always trying to extend their sway to the detriment of the highest education." Nevertheless, he added, "the fact remains that the great boarding-schools were never more popular or more supported than they are today." The public schools had contributed their mite to the war by supplying officers who had bolstered the reputation of the schools, not least by laying down their lives in the conflict. He suggested, moreover, that the public schools were now receiving pupils from middle-class families that were unhappy with the expansion of the local grammar schools and the growing involvement of the State:

> The admission of the "Free Placer" from the elementary schools in considerable numbers to all the great day-schools which have accepted State-aid has caused great numbers of middle-class parents to look elsewhere for the education of their sons, and it is not too much to say that from every city and large town in the country are sent recruits to the boarding-schools who, a generation ago, would have been sent quite contentedly by their parents to the great day-school of the locality.

According to Norwood, this was likely to lead to further growth of the public schools, and to the foundation of new schools of this type.

He argued that this new expansion was not entirely beneficial for the education that they provided, especially as it had encouraged the development of the Common Entrance Examination, "that stiff fence which the boy of 13+ must take in order to reach the coveted land." In Norwood's view, the effect of the Common Entrance Examination was completely deleterious: "There is nothing in English education at the present moment the influence of which is more generally admitted to be harmful than that of this examination, and unfortunately nothing which it is more difficult to reform."[5] On the other hand, he reflected, the public schools had—over the past generation—undergone reforms that were generally progressive in nature: reforms such as a broadening of their curriculum, their participation in the School Certificate examination system, an increasing role for music, art and drama, and a more diverse and less obsessive approach to games.

Another short piece by Norwood, this time in the weekly magazine *The Spectator*, drew out the contribution of the public schools to social service. The General Strike of 1926 had demonstrated the increasing political and industrial conflicts of English society, and Norwood was anxious to present a perspective of the public schools that showed their capacity to bring down rather than to reinforce social class barriers. He acknowledged that the public schools themselves were a significant factor in the growing segregation of the social classes. Nonetheless, he argued that the products of the public schools had goodwill and sympathy for others, even if they were not always at ease with them. Social estrangement, he insisted, was "undesired and accidental."[6] According to Norwood, it could be reduced if the public schools could emphasize the duty of social service, not only through school missions that had been fashionable at the end of the nineteenth century, but also through a range of different initiatives designed to establish contacts between the social classes. In Norwood's view, the aim of such contacts was not to bring charitable relief to those who were "down and out," but rather "to establish chances of mutual understanding between two classes which are each of vital importance to the country."[7] This could best be achieved, he opined, through clubs and camps, especially the latter, and in combination with each other, so that "the ties which are easily formed in the unconventional camp-life can be strengthened in the more permanent quarters of the club by occasional visits." He emphasized that it would be the public school boys and old boys who would provide the leadership in such social encounters: "He might manage the games of his factory or his engineering works; he might hold a Territorial commission, and take some interest in his men."[8] Norwood argued that this kind of initiative was more likely to be successful than sending the sons of poor families to public schools. After all, he reasoned, the elementary school curriculum did not prepare the pupils of these schools for the public school, while secondary schools were not likely to give up their own pupils in order for them to go to a public school and so would not willingly cooperate with such a policy.

A more ambitious and extended article on boys' boarding schools, this time in an edited collection about the different types of schools in England, dealt at greater length with the problems of the public schools and also considered possible solutions to them. He reiterated his view that the origins of the public schools were in the Middle Ages, specifically with William of Wykeham and the founding of Winchester College in 1382. These, he held, had established a common life, a common ideal of character formation, a tradition of sound learning, a close connection with Oxford and Cambridge, and opportunity for all social classes as their key characteristics. Once again, Thomas Arnold emerges as the hero who revived the ideals of the public schools in the nineteenth century. As in his previous contributions of this type, too, Norwood castigated the influence of the Common Entrance Examination. Here, he suggested as a reform that the written test should be abolished, and that the curriculum of the top forms of the preparatory schools should overlap with that of the bottom forms of the boarding schools. The sole test for moving from the preparatory school to the public school, he proposed, should be oral, through an interview. There is also an interesting passage on the public schoolmasters who have remained at the same school throughout their teaching careers, on the "intellectual deadness" that this can produce. According to Norwood,

> Masters have in some cases known nothing else save for a brief interlude at Oxford or Cambridge. The system they have known satisfies them, and they do not know what is going on outside, because they neither read nor visit. These grow into the second-rate obstructives of the boarding schools, of whom there are still too many. They regard themselves as the custodians of the most valuable traditions of the past, and their ignorance of the rest of the national system of education can be gathered from the fact that they still think that a secondary school is a kind of what they still know as a "board school."[9]

He added that such masters were declining in number and their influence had reduced, but there is more than a faint echo in these remarks of his own bitter conflict with masters such as C.G. Pope.

Norwood also dealt at greater length in this article with the role of the public schools in the segregation of the social classes. He repeated his view that the free-place system introduced in the grammar schools, by enhancing a mixture of social classes in these day schools, had encouraged an exodus of pupils from wealthier families into the public schools. This had led to a clear cut separation: "The independent schools, which are almost wholly boarding schools, are now segregated as concerns both masters and boys, and the two sections of the nation, which need to work together throughout their adult lives for the general good, are being educated in isolation from each other."[10] In this situation, he argued, it was necessary to impose a scheme on the schools "to break down the cast-iron barrier of segregation which now divides type from type."[11] This would involve the boarding schools offering free tuition for boys "of strong character and physique" from the day schools, selected by interview on the basis of their school record, with the State paying their boarding fee together with a bursary. Such a scheme would need to be introduced by the Board of Education, with government support, but he argued that it would soon be accepted and would create no friction. Only in this way, he insisted, would it be possible to "break down the barrier of moneyed privilege which now fences off the boarding school."[12]

Norwood's determination to break down class barriers may appear surprising for a generally conservative thinker, but this was a highly characteristic argument in at least two respects. First, it looked to the State to resolve social class differences. Norwood's ideas here echoed those of Matthew Arnold half a century before, when Arnold had depicted the State as the "best self" and aspired to unite the schools of the middle classes under its auspices. Second, it was impatient with entrenched interests and established structures. This latter tendency made him an interesting and popular writer and speaker. However, his irreverent and unorthodox style of public debate also ran the serious risk of making enemies in high places.

This iconoclastic aspect of Norwood's educational thought was also evident in a short survey of the English educational system that he produced in 1928. His approach was markedly different from that of Sir Lewis Selby-Bigge, a former permanent secretary at the Board of Education, who ventured into print the previous year with his reflections on the role of the board. Selby-Bigge confessed his "trepidation" in writing about controversial subjects, especially since, as he put it, "comment upon them may lead to undesirable inferences as to the attitude of the Central Authority and the workings of the official mind." Indeed, Selby-Bigge continued, "I sincerely hope that anything which is found amiss may be attributed entirely to myself and not to the Department in which I have served."[13] No such compunction was visible in Norwood's account.

Norwood's study of the educational system emphasized its historical development over the previous century. He suggested at the outset that the system was not logical and did not have symmetry, but that it was not haphazard either. It had "grown from practical needs," and was now "indissolubly bound up with the national life." Indeed, Norwood continued, "It cannot be understood apart from the national history, for it is the product of the national character."[14] He also pointed out particular features of the English system that in his view marked it out from those of other countries. For example, he noted, the English system was practical rather than theoretical. Moreover, it was "inspired from above," in that its ideals had come mainly from the universities rather than being built upward from the elementary schools. Its history had meant that education in England was "a landscape of peaks and valleys rather than that of a uniform tableland," but over the past three decades further development had worked "not to lower the peaks, but to raise the general level of the valleys in the hope that the inequalities will disappear."[15]

Norwood went on to outline the expansion of the educational system as a whole over the past century, emphasizing the significance of the Education Act of 1902, before turning his attention to secondary education. As usual, he stressed the work of Thomas Arnold at Rugby in establishing the model for other public schools to follow. He also defended the continuing existence of a group of independent schools such as these alongside the state system: "In a great State system of education it is probably vital to its health that there should be a group of schools which stand outside it, where experiment can be tried, and personality find expression."[16] He described the public schools as "the most individual institution of all that this country has created in education, in their merits and their faults the most English."[17] They had contributed much to national education, but, he added, "they are confined to the children of the well-to-do, and neither masters nor boys have sufficient contact with the rest of the national system; for it is clear that social prestige can become snobbery, and isolation can become exclusiveness, and segregation can establish caste."[18] In more general terms, he argued that secondary education had seen "a great and very rapid advance" since the turn of the century,[19] although he also recognized that many young people were not benefiting from any education at all from the age of 14. The next step, he proposed, should be not to create a scheme of part-time education for such youths, but rather "to secure a firm hold on all the children from the age of eleven to fifteen, and to make this definitely a course of secondary education, the characteristics of which shall be elasticity and variety."[20]

Norwood took the opportunity to include in this book his now familiar criticisms of the education of girls for failing to "suit the needs of girls as such, the future wives and mothers of the men of the nation."[21] He ranged broadly across diverse topics from technical education to the universities, from the relationship between the Board of Education and the local education authorities to the health and physique of the pupils. Characteristically, too, he concluded the work with a discussion on the importance of teachers, and the development of a teaching profession with a sense of unity and "the fullest measure of freedom, and the amplest room for initiative and experiment that can be conceded."[22] It was this development, he urged, that was most fundamental for the establishment of "an educated democracy such as the world has not yet seen."[23]

Thus, in the later 1920s, Norwood built up a corpus of published material in which he rehearsed his ideas about education for a broad audience. This was clearly intended to defend the position of the public schools and to justify their continuing prominence, and indeed their dominance, in the context of a rapidly developing educational system and changing society. *The English Tradition of Education* was to be different in kind and scope, not simply because of its length but because it attempted to articulate a basic underlying principle about education in England that was much more ambitious than anything he had produced hitherto.

An English Tradition

In *The English Tradition of Education*, Norwood provided first an elaborated discussion of the ideals underlying his notion of the English tradition. He then expounded, again at length, on the different kinds of schools as they then existed, and on the dangers that threatened its future. Finally, he reflected on a wide range of social and political questions and their bearing on educational issues. The work as a whole expressed idealism and optimism, and it was described by Norwood himself as "a statement of faith and hope."[24] At the same time, it also emphasized the threats to these ideals in a way that reveals much about the brittle and insecure character of his vision of the English tradition.

The key argument of the work, as Norwood described it, was a simple one: "It is merely that we have a great national tradition of education, and that we are in danger of neglecting it, of permitting its structure to be impaired, and its foundations to be shaken."[25] This, he held, was not based on any logical theory or created by any particular reformer, but it had "grown out of the life of the nation," and was "taken to be something which everybody knows."[26] It had carried the country through World War I, and also underlay the ideals of the British Empire. It was not the monopoly of any single social class but rather, in Norwood's view, "the common inheritance of all English schools that are free to live a life of their own, and it is steadily spreading from school to school."[27] It was this that held the potential to prevent international war and social war, by engendering common values and standards.

Norwood's notion of the English tradition of education, as developed in this work, was based in ideals of knighthood and chivalry. According to Norwood, Alfred the Great was the founder of this national tradition, establishing the key principles "that the State or the Church has the right to the service of the best brains of the community, and that the whole body of the people, and not merely one particular class, ought to have a common culture and a common outlook."[28] It is the realization of these ideals through the development of the educational system that Norwood identifies as the key significance of English educational history, beginning with the founding of Winchester College by William of Wykeham in 1382. From this point he begins his now familiar story of the decline of this tradition in the seventeenth and eighteenth centuries, and its revival under the influence of Thomas Arnold in the nineteenth. Arnold, he claimed, was "a great commanding officer," although "to the average boy in his school he must have appeared remote."[29] His ideas were embraced by other masters at other schools, so that over the course of a century the old ideals were recaptured: "The ideal of chivalry which inspired the knighthood of mediaeval days, the ideal of training for the service of the community, which inspired the greatest of the men who founded schools for their own day and for posterity, have been combined in the tradition of English education which holds the field to-day."[30] It was based on religion, emphasized games, developed an intellectual appeal, and was inspired by the duty of service. This general tradition was not the monopoly of a group of exclusive schools, he insisted, and he argued that it should now spread quickly "to all the schools of the country."[31]

Having idealized the national tradition of education as a whole, Norwood went on to celebrate its component parts. The most important of these, for Norwood, was religion, the key purpose of which was the production of character.[32] To emphasize this, he devoted three chapters to an exploration of the role of religion. He then expounded at length on the importance of discipline, culture, athletics, and service as aspects of the English tradition. These were closely related to each other, so that, in Norwood's view, the value of games, for example, was in implanting the right ideals of character and conduct. He expressed these in terms that were extravagant even by his own standards:

> They are these, that a game is to be played for the game's sake, and that it matters not a button whether it is won or lost, so long as both sides play their best; that no unfair advantage of any sort can ever be taken, and that within those rules no mercy is to be expected, or accepted, or shown by either side: that the lesson to be learned by each individual is the subordination of self in order that he may render his best service as the member of a team in which he relies upon all the rest, and all the rest rely upon him: that finally, never on any account must he show the white feather.[33]

Similarly, the ideal of service was a further dimension of religion, in that it constituted "no more than the love of God and the love of our neighbour, long ago set forth in that straightforward document, the Church Catechism."[34]

Norwood's account amounted to a particular notion of English national character, and how it was expressed in and through its schools, that was in some ways reminiscent of Matthew Arnold's critiques written in the 1860s. The view of national character that arises from Norwood's work is patriarchal in the sense that it draws on male rather than female images of the English tradition, for example, in athletics and service. It made no attempt to include the education of girls within this vision.

Moreover, it was class-based in its glorification of the English gentleman intent on a mission on behalf of civilization. In *Culture and Anarchy*, Arnold had also stressed the importance of cultural ideals emanating from figures such as Abelard in the Middle Ages contending with the forces of anarchy. Taking its cue from such accounts, Norwood's approach to English history romanticized particular traits and ideals and viewed historical developments over the longer term as the struggle to realize such ideals as perfectly and as fully as possible. This was depicted as an evolutionary process, but one in which the basic characteristics were present from the outset. Thus, he argued, the institutions of English education, like those of the English political system, changed and adapted themselves but maintained their identity from century to century.[35] There was an underlying continuity that was based on a constant inspiration or ideal—and a unique destiny beckoned, should the tradition be realized to its fullest extent. This notion of history was a whig interpretation in the sense that the historian Herbert Butterfield was soon to criticize, a liberal-progressive view of the triumph of one party against the forces of darkness. Butterfield attacked in particular "the tendency in many historians to write on the side of Protestants and Whigs, to praise revolutions provided they have been successful, to emphasise certain principles of progress in the past and to produce a story which is the ratification if not the glorification of the present."[36] This tendency, which, as Butterfield proposed, led to a scheme of general history that demonstrated the workings of an obvious principle of progress converging upon the present, is well exemplified in Norwood's work.

Norwood's book proceeded to locate the major categories of schools in England, the secondary boarding schools, the secondary day schools, and the elementary schools, in relation to the ideals of the English tradition. The boarding schools were popular, he maintained, although not always for reasons that he would prefer. In particular, as he already commented in his earlier essays, class distinctions or snobbery had led some parents to transfer their sons from the local day school to a boarding school. He criticized the schools themselves and repeated his complaints about public schoolmasters who knew of no other ways of life, the effects of the Common Entrance Examination, and the excesses of athleticism.

Norwood criticized the secondary day schools as well, but he emphasized their development since 1902 and argued for the further undermining of the social distinctions between them and the boarding schools. For example, he proposed that the Headmasters' Conference—to which the headmasters of public schools belonged—had become a symbol of class distinction and that it should now be disbanded: "Probably one of the best things that could happen for the future of English education would be that the Headmasters' Conference should cease to exist, and reappear as a Boarding School Committee of the Headmasters' Association."[37] More broadly, he was hopeful that the older day schools might be able to help to "bind the boarding-school system and the system of the State-aided municipal and county schools into a whole, which shall be animated by a single spirit."[38] Thus, the secondary day schools remained central to his visions of how the English tradition might be regenerated for the future.

The elementary schools were also included in Norwood's vision of how the English tradition might develop further. The growth of elementary education was viewed by Norwood as a "steadily civilising agency," and indeed "the main influence which has prevented Bolshevism, Communism, and theories of revolt and destruction from obtaining any real hold upon the people of this country."[39] Norwood argued that religious instruction had been well established in the elementary schools, and discipline improved. Moreover, he suggested, their cultural dimension had been enhanced specifically because examinations had become less pervasive. Athletics still needed to be further developed, while service was also not clearly evident, and the children required something presented that was "more direct, immediate and concrete."[40] There were therefore, in Norwood's view, substantial foundations on which to build.

The implications of this position for future development, in Norwood's view, were that elementary education should end at the age of 11 or 12, and secondary education should then begin for all pupils. In order to achieve this, the school-leaving age would need to be raised by one year. For most pupils, secondary education would be about "the production of sound character more than intellectual performance."[41] Not all children were suited to a secondary education in the full academic sense, and the blackcoated professions were already overcrowded while there was "an unwillingness to enter the life of handicraft and production."[42] From 11 to 15, the course to be followed would be designed to produce "a good physique, practical ability to be developed by more extended teaching of handicraft, a knowledge of scientific method, a good standard of English in speech and writing, and those practical virtues of integrity and the sense of responsibility which a good education can certainly develop." From this course, all those with above average intellectual ability would be drafted "to other types of school, which will follow the old well-ascertained lines." This was a sharp delineation on an intellectual basis, and he insisted that while the nation would need the services of all its "first-class brains," it did not want and could not employ "a plethora of second-class brains which have been attempting a programme which it takes first-class ability to fill."[43]

Such was the prospect that Norwood represented as realizing the English tradition, but he also outlined in graphic terms the dangers that he saw in the path. These were in particular mechanization, misrepresentation, and individualism. Mechanization involved according to Norwood "the cramping and deadening influences which may proceed from the too rigid working of the necessary machinery which is external to the life of the schools, and merely intended to provide the conditions that are most favourable to growth."[44] He acknowledged that this tendency might most readily be associated with the Board of Education, but he insisted that this was no longer fair as the board now tried to avoid prescribing in too much detail. It was the local education authorities on which Norwood cast suspicion for a frequent tendency to insist on uniformity. External examinations were also potentially dangerous to the life and freedom of the schools, in Norwood's view, especially where they were used to impose the curriculum. Rather, he proposed, "Up to the age of fifteen a child should not be examined save by his or her teachers—this in the interests of sound teaching: one does not pull up the tender plant to examine its root formation."[45] From this point, there would be an examination system designed for pupils who were fit for universities, professions, and the higher posts of industry, manufacture, and commerce. Thus, for the mass of pupils there would be no examinations at all, instead there would be "testing and experiment" to discover the form of education that was best suited for children of limited intellectual ability.[46]

The misrepresentation of the English tradition was a further danger against which Norwood gave warning in his book. Such calumny included the argument that the English tradition was a class ideal rather than a national ideal, or that public schools were out of date and irrelevant to the modern world. He singled out the ideas of Bertrand Russell, a distinguished philosopher of education, as a prime example of an unfair line of attack on the old educational tradition. In this context, he went out of his way to defend the contribution made by the public schools to the expansion of the empire, especially in India, which the public school products made "a country where men can walk in peace, and trade can flourish, and justice be done, and they so bore themselves that many won the devotion of the 'inferior' race: if they fell short of this, it was because they belied the tradition of their bringing-up, not because they kept it."[47] Here again is an intriguing echo of his father Samuel with his lavish praise for "our Indian Empire."

The third and most serious danger to the future of the English tradition in education, in Norwood's opinion, was posed by individualism. Bertrand Russell's recent book *On Education* was presented as a key example of the fallacies of an individualist approach.[48] Norwood was insistent that giving priority to the individual as an end in itself was not only an insubstantial and self-defeating ideal, but that it also undermined and threatened the English tradition and more broadly the essence of western culture. It was far preferable, he persisted, "to teach the child not to seek the self-realisation of the individual self, but to live as a member of a community, to be loyal to its laws, to seek to remove its injunctions and its imperfections, and to deserve well of his fellow-men."[49] He set his store rather by the ideal of service as the essential means by which to foster the democracy of the future.

The remainder of Norwood's book was devoted to an assessment of how contemporary developments in politics, society, and international relations would impinge on the character of education. He dwelt first on the role of knowledge in democracy and suggested that a new era of "control and security" might be established based on knowledge and its application. This, he proposed, would entail "handing over life to the guidance of those who know, not to those who are amateurs at improvisation, however brilliant, or to those who with stoical perseverance under hard knocks are content to muddle through."[50] Cooperation under the direction of educated experts would lead to improved health and more efficient production. At the same time, all citizens would have gone through an education that, even if diverse in content, would be similar in ideals. These precepts, according to Norwood, would underpin a viable democratic alternative to the fascist mentality of Mussolini in Italy and the communist system of the Soviet Union. In relation to the new challenges facing international cooperation, Norwood argued the British Empire remained an important influence for the promotion of peace, while the League of Nations offered new hope for international cooperation. He endorsed the Treaty of Washington of 1921, and the Treaty of Locarno of 1925, as key instruments for the prevention of further war and suggested that the schools should emphasize the history of international relations over the past century to highlight the future challenges.

Finally, Norwood appealed directly to the middle classes as his principal audience and concern. Amid the rapid industrialization of the preceding century, he noted, the individual was tending to suffer and had less political influence than hitherto.

The middle classes had fallen victim to the processes of industrialism: "The class that should produce men and women of sweetness and light, lies crushed between the upper and nether millstones of capital and labour, the sole real forces necessary to the operation of the system."[51] The Liberals were dwindling, while Conservatism represented capital and Socialism represented labor. He argued that while socialism was becoming outdated, and class war irrelevant, the industrial system itself tended to treat individuals as little more than economic units. In order to promote culture and individuality in these circumstances, Norwood suggested, support should be given to groupings to oppose the influence of mass civilization, especially by enhancing public taste, maintaining freedom of speech, encouraging a broad intellectual expertise, stressing individuality in education, and uniting the teaching profession in a common ideal. The family and the home, he added, should also be safeguarded against the effects of the industrial system. His book concluded with the lofty sentiments that the future should not be one of enslavement to machinery, nor one in which millions live in poverty while a few rise to a purer air, and that both prospects could be averted through an education that was "fully conscious of the perils which beset modern humanity, and places spiritual values first for all men of all classes."[52]

The English Tradition of Education was thus an ambitious and expansive work. It ranged widely across the problems facing contemporary English education and attempted to locate these within a rapidly changing social, cultural, and political landscape. It provided historical depth and also comparative and international insights to inform the issues of the day. It was clearly and often engagingly written, not simply for specialists in education but for a wider readership. It was frank and unapologetic, often reveling in iconoclasm. Yet Norwood's book had definite faults and limitations. At heart, it was a deeply conservative work. For all its liberal values, it emphasized elitist solutions. The intellectual underpinnings of the work were those of Plato, and while its principal hero was Thomas Arnold, it was solidly in the tradition of Matthew Arnold's *Culture and Anarchy*. Moreover, despite its breadth of scholarship, it lacked originality and depth of analysis. As a contribution to educational debate, too, it was open to criticism. Although it emphasized the needs of the future, it remained rooted in the past. It tended also to be politically ambivalent, too conservative in its overall stance to appeal to reformers, but too fond of critical comments about established interests and institutions to win the warm approbation of conservatives.

Responses and Reviews

Norwood's book did attract a wide range of responses and reviews, and these revealed a mixture of admiration and skepticism. The ideal of secondary education that the work articulated was given a broadly respectful and sympathetic reception from many contemporaries who were only too pleased to see such a sturdy defense of established values combined with interesting commentary on the pressing issues of the day. Nevertheless, it also provoked critical comments, on the one hand, from unreconstructed apologists of the public schools, and on the other, from proponents of public school abolition and a more radical reconstruction of the world of secondary education.

Among Norwood's closest acolytes, unsurprisingly, there was lavish praise for the sentiments voiced in his book. Among the first to write was Ronald Gurner, a former assistant master at Marlborough College and now headmaster at Whitgift Grammar School in Croydon. Gurner was fervent in his support:

> I read your book today: well done! You know how for many years some of us have taken without question your general lead in educational matters, even if we have been foolish enough to disagree in points of detail? So this new book of yours simply gives us a new text book, which I for my part shall use without scruple.[53]

He added for good measure that he would take the book to a local meeting of the National Union of Teachers to spread its ideas about teachers' associations. T.F. Coade, a young assistant master at Harrow, was also quick to give his approval: "It does not require much imagination to realise what thought and experience must have gone to the making of such a book, and the clarity and grace of style must make it good reading even to those not primarily interested in education."[54] Another admirer, Dr. G.F. Morton, the head of Leeds Boys' Modern School, added his ready assent: "The feeling that is dominant in me for the moment is one of immense gratitude that you have written a book which, coming from the Head Master of Harrow, cannot fail to have a vital influence not merely on educational but on world issues."[55]

Beyond Norwood's immediate circle, too, much attention was given to Norwood's book. Such was the strength of interest, indeed, that the publisher John Murray soon decided to order a reprint, adding in a letter to the author: "I was dining the other night with the Dean of St Paul's [W.R. Inge] who was enthusiastic about the book ... on all sides I hear similar approbation which is pleasing in that it shows that really good books are sometimes properly appreciated!"[56] This response must have been welcome to Norwood, especially coming as it did in the final stage of his struggle with C.G. Pope at Harrow. Nonetheless, there were also strong criticisms that were a reminder of the antagonisms that Norwood attracted, including one from a brigadier who criticized him in no uncertain terms: "From your English Tradition of Education you condemn yourself as a whole-hearted liar—an ambitious snob—and one of the worst specimens of half-educated headmasters that any public school has endured. *Leave Harrow alone* .... Someone will make the position too hot for YOU to hold ... I speak for several sons of Old Harrovians."[57]

There were many reviews of the book published in specialist educational journals and more general periodicals and newspapers, and a large number were unstinting in their praise. C.W. Valentine, an educational psychologist, was one of the most lavish in this respect, writing in the *Forum of Education* that Norwood's work was "extremely lucid and readable," brave, remarkably impartial, and marked by great breadth. According to Valentine, indeed, "this is one of the most important books on education of the present century."[58] Another review, by Lord Gorell in the *Quarterly Review*, described it as a "noble book, rich in the high ideals and the practical experience of one of the foremost protagonists in the field of English education today."[59] Frederick J. Mathias, in the *Western Mail*, ventured to propose that Norwood's book was "the most vital utterance on education that we have had for many years," so much so that it was "too precious to borrow," and "should be the personal possession of all teachers, directors of education, members of education committees, and incompetent and unqualified critics throughout the country." Mathias concluded, "This is a book for Everyman, his family, and his political representatives. It is a new Declaration of Educational Rights."[60] Further support, although not always quite so effusive, came from periodicals as far removed from Norwood's own religious views as *The Tablet*, as different in its politics as the *New Statesman*, and as geographically distant as the Johannesburg *Sunday Times* in South Africa and the *Otago Daily Times* in New Zealand.[61]

A number of other reviews, while generally sympathetic, took the opportunity to offer criticisms of the tradition as depicted by Norwood. For example, a review in the *Times Literary Supplement* was both kind to the author and shrewd in discussing the issues raised by his book. According to this review, Norwood enjoyed an exceptional stature in the educational world that went beyond his current position at Harrow:

> By his ability and commanding personality he has acquired an exceptional influence with schoolmasters, governing bodies, the Board of Education, educational associations—with all, in fact, who are interested in education, particularly secondary education; and the fact that the first stage of his remarkable career was the very successful headmastership of Bristol Grammar School has given him a position in the esteem of the great body of teachers throughout the country which it is safe to say no other headmaster of one of the great boarding schools has ever enjoyed.[62]

It suggested that the "immense improvement" in secondary education since the 1902 Act, together with "a certain mellowing of the author's mind which is the natural result of his successful career," had made Norwood less impatient than he had been earlier in his career. Nevertheless, it added, his characteristic qualities remained intact: "Dr Norwood is on the whole conservative and suspicious of innovation. He has rather a short way with views which do not commend themselves to him. He has the great merit of never leaving the reader in doubt about his meaning or his opinions, whenever he is, as he usually is, quite sure of them himself." At the same time, it noted that his emphasis on religion led him to understate the importance of other dimensions such as duty and patriotism. Also, it suggested, his comments on current dangers lacked any sense of compromise or balance, so that "here, as elsewhere, one is left with the impression that the Roman and the practical man are so strong in him that he cannot take the first step towards converting his opponents, that of understanding them."[63] These were shrewd criticisms of Norwood's style of debate that often alienated rather than persuaded others.

A similar note was struck in a further review of Norwood's work published in *The Times*. This also remarked that his book was "well-written, well-printed, handy, and handsome," and that it would "meet with the attention due both to his position as Headmaster of Harrow and, even more, to the remarkable reputation which he has won by his personal gifts and achievements."[64] However, like the review in the *Times Literary Supplement*, it observed that Norwood's style of argument was as likely to stimulate opposition as it was to win friends: "while he has a manner which is perhaps even more judicial than magisterial and presents his arguments with studied sobriety and marshalled evidence, there is a certain asperity in his style and a certain impatience of opposition which are more likely perhaps rather to hearten his supporters than to conciliate the unconverted." It also pointed out that while Norwood was basically conservative in his approach, he was also unconventional and radical, so that "the more conservatively minded of our educationists look to him as a champion somewhat in the way in which the political conservatives of Rome looked to Pompey and those of Victorian England to Disraeli." It added that while Norwood was a statesman of the educational world, "perhaps for that very reason his book has not quite that touch of inspiration which would make one inclined to hail him as a prophet or sage," concluding that although it should stimulate thought, it would "hardly become a classic on its subject."[65]

One particularly distinguished reviewer of Norwood's book was Sir Michael Sadler, who had played a major role in the developments that led to the Education Act of 1902. He also voiced admiration for Norwood's own personal qualities and experience and argued that his book was of "unusual importance." Indeed, according to Sadler, "Years hence this book will show what were the hopes and fears of the headmaster of a great English public school when he reviewed the wide and varied scene of English education and pondered on the future of his country in international affairs." Yet Sadler was skeptical as to whether Norwood would succeed in his self-appointed mission of uniting English education around a single tradition: "Dr Norwood's book leaves the reader doubtful whether it will ever be possible to fuse together the different metals which, for many centuries, have existed in that rather shapeless amalgam which we call English education."[66] This was a significant warning, coming as it did from an observer with a long experience of English education, that Norwood would ultimately fail to find sufficient supporters for his approach to education. His ideal of secondary education, while arresting and elaborately argued, could not heal the educational and social divisions that had become so evident.

It was notable, too, that among the most critical reviewers of Norwood's book were two younger commentators with a keen eye for the social issues that were to be addressed over the next generation. The first was H.C. Dent, a young teacher with progressive inclinations, who was later to become the editor of the *Times Educational Supplement*. Dent argued forcefully that Norwood's vision of an educated democracy based on the English tradition had very little connection with most people outside a few favored public schools. There were, he insisted, two English nations that "live apart from each other, are educated differently, think differently, are, in fact, complete strangers to each other." Therefore, according to Dent, "Not until the educational needs of *both* these two nations are fully met shall we have an educated democracy in England."[67] At present, he continued, the public schools served the interests of the ruling or privileged classes, while the elementary schools and secondary schools were designed for the rest. There was no opportunity for the English tradition of education to flourish in the elementary schools, while the secondary schools admitted only a small minority and tried to imitate the public schools. The educational system was therefore developing in a way that was inappropriate for the needs of the country as a whole, under the guidance of an elite group that had no understanding of these needs or of anything outside of their own limited experience. For this reason, only radical measures would suffice: "education has got to be thought out in England from the very beginning, and right through from A to Z, without prejudice, without longing side-glances at any type of education, however good, now in existence, but quite dispassionately and solely with regard to the facts."[68] He agreed with Norwood that spiritual values should be emphasized and added that political education should be provided in order for schools to become an apprenticeship for public as well as for private life: "You must not forget education for livelihood, nor for leisure, nor for the public life, but you must devise it in terms of the life from which these children start and in terms of the life they may achieve."[69] Overall, this was a trenchant critique of Norwood's ideas and a thoroughgoing challenge to many of his basic assumptions.

Another challenge on broadly similar lines came from Kingsley Martin, who was soon to be appointed as editor of the left-wing weekly periodical the *New Statesman*. Norwood, he suggested, was attempting to answer the critics of public schools without understanding the point of their criticisms. Martin made so bold as to inquire, "What is one to do with such a simple-minded man?"[70] He had himself been through two public schools, receiving education that he held had been an inadequate preparation for the needs of modern society and politics: "To some public school-boys today, it seems unfair that they should have been stuffed up with a childish notion of school as a microcosm of the world, to which 'playing cricket' is the right solution." According to Martin, then, Norwood was too immersed in the public schools to understand their problems, being himself "so much a public school product that he does not really conceive that in any essential the public school system could be wrong."[71]

Thus, Norwood's *English Tradition of Education* developed from his own experiences, personal, professional, and as a leading policy maker, to project both his fully developed ideal of secondary education and a sense of crisis at the dangers that confronted it. The book was a success in that it received wide attention and was generally given respect and at least a measure of support. Nonetheless, it represented only a partial and whiggish view of English educational history. Its lofty prose and the idealized and mythologized tradition that it exalted were also highly inadequate for the task of addressing increased demands for equality for people as a whole, or for technological change or industrial development. The dissenting criticisms of the younger generation struck a jarring note amid the praise and served advance notice that opposition and antagonism were to become more widespread and more sharply defined in the years ahead.

Chapter 8

The New World of Education

The 1930s witnessed the rise of fascist regimes in continental Europe, and a slide into international conflict, and ultimately a second world war. These new threats encouraged educators to find new ways forward as alternatives to fascism that were based on democracy and citizenship. Norwood was closely involved in the suddenly urgent search for a new world of education. He was consistently and strongly antifascist in his outlook, and a supporter of initiatives that were characterized as "progressive." He continued to find himself much in demand during the 1930s as a keynote speaker at conferences, and as a contributor to publications on a range of topics. These activities afforded him a platform for his ideas even after he had left Harrow for the less prominent position at his old Oxford college. This engaged him in a small elite network of intellectuals, teachers, and academics who were predominantly—if not exclusively—male. Moreover, Norwood was now well known internationally as well as within his own country, and he took advantage of this by being actively involved in conferences in different countries. In 1930, he took part in a delegation to Canada to discuss developments in education. Seven years later, he contributed to a lengthy series of conferences under the auspices of the New Education Fellowship (NEF) that took him to the United States, New Zealand, and Australia. These new developments also had an increasing bearing on his relationship with the public schools. In the 1920s and 1930s, he promoted the establishment of new boarding schools included in a network entitled the "Allied Schools." By the end of the 1930s, however, growing economic, political, and social pressures led him to advocate bringing the public schools under the protection and control of the State.

Education for Citizenship

Taking part in conferences and public meetings on a regular basis, and the traveling that this entailed, was a demanding and often exhausting experience. Norwood was well accustomed to this, but as he grew older it increasingly took its toll. He also found it difficult to conceal his impatience with colleagues who were sometimes his traveling companions on long journeys to take part in conferences. One routine engagement is recorded in a January 1935 letter from Norwood to his wife. He had traveled to Liverpool Street train station in London to meet his son-in-law Clifford Canning and the philosopher of education Sir Richard Livingstone. With them he had gone by train to Cambridge, and thence to Clare College, suffering from the cold. He presented a paper in the evening, followed by discussion, dinner, and evensong, before having a long talk with Spencer Leeson. In the morning, he had breakfast with Cyril Bailey, listened to a long paper by Sir Francis Goodenough, and then himself spoke for about twenty minutes, before leaving to travel to Brighton where he stayed at the Norfolk Hotel. Being in the public eye and always likely to stir controversy, his own contribution was quoted in the newspapers: "In spite of there being no reporters I find myself being reported in the Sunday papers."[1]

Opportunities for such public discussion were also heightened during the 1930s through the development of new organizations and informal networks designed to promote new thinking about education. Norwood himself had encouraged the introduction of regular conferences of junior public schoolmasters, organized by T.F. Coade, an assistant master at Harrow and from 1932 the head of Bryanston School, a newly established public school on progressive lines. These conferences were held at Harrow School from 1930 onward and attracted a number of highly distinguished speakers with major contributions published by Cambridge University Press.[2] He also helped to promote the development of a more formalized network, the Association for Education in Citizenship (AEC), established in 1934 by a former Liberal member of parliament, Sir Ernest Simon. Such initiatives gave Norwood ample scope to consolidate his influence and authority, while adapting his key themes to the changing debates of the time.

Longer visits to overseas locations brought further everyday pressures, as well as opportunities of international recognition. In 1927, at a conference in Venice, he discovered that, as he told his wife, "the biggest ass in Wales was of the party, one Edwards, who represented the Principality on the Examinations Council for years—his wife and son complete the Party." Norwood added, "I shall not lack amusement of a quiet sort."[3] On a longer journey, to Canada in 1930, he sailed with 15 other prominent educators on a large ocean liner, the Canadian Pacific's "Duchess of Bedford." As he noted, "It is to my eyes, a huge boat, like a city, and we have miles of restaurants, smoking rooms, lounges, working rooms. But then I have never been on a big liner before." For Catherine's benefit, he confessed that he felt "rather lonely and sad, coming away, and leaving you all alone for so long, or at least being separated from you for so long."[4] Nonetheless, the visit was a resounding success, and he was moved to tell his wife of the wonderful view he had enjoyed of the Canadian Rockies. Every day, he enthused, was remarkable: "I don't think I shall ever be able to pass on the wonderful nature of the experience. At any rate it can never happen again." Meanwhile, he was rewarded by the glowing tributes of his hosts and colleagues: "My vanity was fed last night by the fact that the Chairman said I had been described by one member of the delegation to him as the greatest personal influence that had appeared in English education for many years. It is rather hard to live up to that sort of thing." The involvement in a large delegation—which could be tedious over a long period such as this—was also not a problem on this occasion, as he acknowledged, "All is well socially—there has been no unpleasantness of any sort and I think it will remain so."[5]

Through such activities, Norwood remained visible and active in disseminating his authoritative views on education, views that were at the same time often quixotic. During his visit to Canada in 1930, for example, he had the opportunity to deliver a major address to the Canadian Club of Montreal on "The Education and Outlook of English Youth." In this speech he claimed, very surprisingly on the face of it, that English education was going through a wonderful period of educational progress. He based this claim not on the immediate or current developments in education, which were not promising as the education system struggled to cope with the consequences of a severe economic slump,[6] but on longer-term trends. In particular, he argued, there were two major systems of education, the elite system and the mass system, that coexisted in England and were now being brought into a closer relationship with each other. He suggested that this was taking place gradually through the development of a curriculum that was very similar to the curriculum in public schools and in other secondary schools.

The major challenge that Norwood identified in this address was how to extend the education of the majority of the population without reducing the standards of elite education. Such an extension, he insisted, would not constitute "secondary education for all." Indeed, he brusquely dismissed the phrase that had already been successfully popularized by R.H. Tawney and the Labour Party as "a silly slogan, although we often hear it," and maintained that in England "we have already provided sufficiently for those who are capable of profiting by an ordinary academic education." At the same time, he acknowledged that it was difficult to plan a suitable course for the extended education of the majority of the population. For this purpose, he proposed a greater engagement with activities such as games, music, dance, and drama, to "try to get down to those instincts which are dormant." In general, he concluded, "We have got to throw away a good many prejudices and open our minds very widely and handle our boys and girls from eleven to fifteen in a very different spirit from that of the past."[7] Despite this clear awareness of the problem, however, the details of the curriculum that would be necessary for such a purpose remained very much unresolved.

Amid the gathering international crisis of the 1930s, the major note that Norwood struck in addressing the issue of educational differentiation, and a range of other challenges, was to stress the importance of providing an education that would prepare the citizens of the future. This emphasis on education for citizenship, already a feature of Norwood's thinking earlier in his career, became a dominant theme during this period. Such an approach, he argued, was crucial in underlining the differences between English education and the education system of the fascist states. This was evident, for example, in an address that he delivered at the first conference of junior public schoolmasters at Harrow School in January 1930, attended by about hundred and fifty men. Portentously titled "Unity and Purpose in Education," this address was a clear signal of both the national and international issues involved in education for citizenship. Education shaped itself in his mind, as he explained here, "primarily as an education for citizenship."[8] It needed to respond to the needs of the national life, and to continually change and develop, in order to produce servants of the community with well-developed individual character and gifts. He argued that the worsening national situation, with growing unemployment, industrial unrest, and a general fading of the advantages enjoyed by the country in the nineteenth century, increased the urgency for this kind of approach to education to be developed further. The priority now, he proposed, was for teamwork and unity, the excellence of products, and the development of new ideas: "Our future depends upon the application of science to industry, in other words, it depends upon our education." It was important also to produce leaders who were fitted to administer and govern in a widening range of situations:

> Notably in Africa, but also in Asia and elsewhere, we want not only the cadet of the governing stock of whom in the nineteenth century we produced sufficient, but also scientists, entomologists, directors of agriculture, doctors, and engineers of many types, who are all capable of research, or of applying the results of research, and capable also of governing, teaching and directing native and backward populations.[9]

For the colonies, therefore, the demands of education for citizenship apparently entailed new forms of leadership provided by the British, rather than independence for the people themselves.

Within Britain, meanwhile, the problem apparently was somewhat different in Norwood's view. The need here was to improve education for the mass of the people so that they would better understand the problems of the modern world. Already, at the beginning of 1930, he was conscious of the potential threat posed by the rise of fascism, and the need to develop an appropriate alternative form of education. The system being developed in Italy by the fascist dictator Benito Mussolini, he declared, meant in effect "to treat the greater part of the nation like cattle, who have to be made to do what is best for them, since they are incapable of judging for themselves."[10] Norwood insisted that such an outlook was totally alien to the English character, although clearly there were significant tensions in his own approach. While endorsing the need for the majority of the population to be able to judge for themselves, he continued to proselytize the kind of messages that he considered most appropriate for them. Moreover, he remained unwilling to allow the indigenous populations of the British Empire even this limited form of discretion.

At the same time, Norwood also subtly recast his religious ideals to take account of deteriorating national and international prospects. He perceived the international crisis in terms of a moral crisis of organized hostility to Christ and Christian knowledge, in which materialism and economic values took precedence over spirituality: "What shall a man gain if he gain the whole world and lose his own soul? This, then, for me is the fundamental question of education, to guide the young to realise first that this choice has to be made by all that are born to this inheritance of human life, and then to realise what is the right choice."[11] Therefore, he argued, it was more vital than ever to insist on "the supreme importance of spiritual values," to go "through all the schools of our country, through and through our national education."[12] This would require repentance to be inculcated in all citizens through education. Norwood acknowledged that this was increasingly necessary for the United States of America, in order for it to assert a moral approach to its increasingly powerful position in the world, but he insisted that it was still Britain that had the most important strategic role for the future, "for we may still lead the world, and if we can, render our greatest service to humanity, greater even than liberty and freedom of thought and of speech and parliamentary government, of which we have been the pioneers."[13] This idealized vision of a national history, character, and destiny, closely linked to basic assumptions about Britain's imperial duties and wider responsibilities in a changing world, thus continued to underpin Norwood's broader ideals.

Norwood also continued to champion an enhanced role for the State, while recognizing the anxieties of the public schools and the rival claims of freedom and independence from State control. These issues were addressed directly in a fascinating lecture that he presented in the Kingsgate Chapel in London in May 1932, the ninth Shaftesbury lecture, under the title of "Scylla and Charybdis, or Laissez-Faire and Paternal Government." The lecture was in honor of the seventh Earl of Shaftesbury (1801–1885), who had led a campaign against child labor in mines and factories and chaired the Ragged Schools Union for over forty years. Norwood's discussion of Lord Shaftesbury as an exemplar might also be read as an oblique reference to his own experiences. He pointed out that Shaftesbury had distrusted pleasures and the amenities of life because of the effects of his childhood, when his mother neglected the family, and he became alienated from his father: "In his boyhood his father was a person to be feared, and in his manhood neither friend nor counsellor, but estranged and often hostile."[14] Moreover, according to Norwood, although Shaftesbury lived to be praised and honored, "he might well have been a disregarded prophet, and in all *a priori* considerations might have been expected to do so, for his whole attitude of mind was foreign to the prevailing spirit of his age, and he disapproved of the whole main stream of the development of his period."[15] He objected to the philosophy of "laissez-faire," in which the market rather than the State would be left to decide on educational and social provision. Yet he was equally averse to democratic principles such as "government of the people by the people for the people," as proposed by his great American contemporary, Abraham Lincoln: "What he wanted was government of the people for the people by those who knew best and cared most."[16] This was a fundamentally elitist conception of government that led Shaftesbury to oppose many of the major social and political reforms of his time. Norwood clearly recognized a kindred spirit in both personal and policy terms. He also made use of this opportunity to emphasize the problems of extremes in both "laissez-faire" and paternal government, or, as he described it, "the problem of too much direction or too little."[17] It was necessary, Norwood insisted, to find a balance or a compromise between them: "Somewhere between Laissez-faire and Paternal Government there lies a happy mean."[18]

Norwood's religious, social, and political views located him close to Liberal Anglicans such as William Temple, the archbishop of Canterbury, Ernest Barker, a leading political scientist, and A.D. Lindsay, the master of Balliol College Oxford. Like these other interwar English thinkers, he emphasized the ideal of community and articulated it as an alternative to communism and Nazism. According to a recent study by Matthew Grimley, Liberal Anglicans believed in an organic national community (developed as the realization of a plan involving both the Church and the State), were opposed to economic competition and social class conflict because these undermined national unity, and looked to education as an influence to counter them.[19] They were also influenced by the ideas of Thomas and Matthew Arnold, and by the Idealism of T.H. Green. Along with these others, too, Norwood's continued influence helps to demonstrate that religious and specifically Anglican ideas maintained a key role in educational and social development up to World War II. Yet Norwood retained an elitist and a nationalistic emphasis in his thinking that tended to mark him out from such contemporaries. These aspects of his approach also made him an unusual and in many ways an uncomfortable recruit for the new crusades of the 1930s represented by the Association for Education in Citizenship and the New Education Fellowship.

Norwood developed his ideas on education for citizenship in more detail and emphasized their implications for the school curriculum in a presidential address to the Science Masters' Association at the end of 1931. That he was invited to become the president of the Science Masters' Association was in itself remarkable in view of his general indifference to the subject, and the his disinclination to its further development at Harrow. On the other hand, it did reflect his high status and the extent of his influence at this time, as well as the breadth of his own scholarship. His presidential address was characteristic in its combative and uncompromising approach, and also in its interest for the linkages that it made between science and citizenship. He began it with a trenchant analysis of the crisis that confronted modern civilization:

> The general position of the world to-day, both in theory and practice, is as that of some bather advancing into ever-deepening water, who feels the last slope of the solid bottom slipping away from beneath his feet, and is conscious that he must shortly swim for it, in the hope of reaching solid ground on the other side, though he is not sure that there is another side.[20]

This crisis, he held, was partly political in character, in that authority of all kinds, including scientific reason and religious doctrine, was coming increasingly into question.

In international relations, beginning with World War I, there was a movement based in primitive passions rather than reason that increased the threat of renewed conflict. Therefore, a recourse to scientific method was according to Norwood a fundamental means of resisting further warfare and of building for the future. On the other hand, he suggested, it would be unwise to leave new developments in science and industry to businessmen and practical people such as the American industrialist Henry Ford, because this would leave too much power in the hands of individuals who might not use it in the public interest. Indeed, Norwood argued,

> It is possible to imagine immense powers of destruction resident in some poisonous gas or explosive placed in the hands of some ambitious and unscrupulous statesman, who knew how to exploit the animosities of his nation, and who would have the power to destroy a capital city, and all the treasure of civilisation and human achievement that was accumulated within it.[21]

In these circumstances, Norwood recalled the ideas of Plato for his ideal republic in which a class of guardians would organize and protect society. His disbelief in the selective breeding of such leaders, or eugenics, also set him apart from a fashionable nostrum of the early decades of the century.[22] It was, rather, education that he looked to for such an outcome, especially if the character of education could be developed further to promote an informed understanding of society and politics. Schools would then become not primarily a "rather exclusive training-ground for the professions," but instead "the training-ground for democracy."[23] Science would have an essential place in the curriculum, not so much in the production of specialist chemists or physicists but in "laying the foundations of principle for a future citizen."[24]

The Spens Committee on Secondary Education

It was with these ideas influencing elite opinion formers that Norwood became actively involved in policy discussions of how to promote education for citizenship in the secondary schools. He still had friends in high places, among them Edward Lyttelton, the former head of Eton College. In May 1933, Lyttelton contacted the president of the Board of Education, Lord Irwin, with some ideas about how to "establish order in the chaos of Secondary Education." He was careful to note that he had been in discussions with the "very capable" Norwood, while taking responsibility himself for the views that he put forward, on the grounds that Norwood's "official entanglements" made it difficult for him to give open support to an "additional crusade."[25] Lyttelton proposed that new approaches to the secondary school curriculum were urgently needed to counter the effects of the growing influence of examinations and economic priorities. A more rational curriculum might be developed by a small committee established to agree on fundamentals, a curriculum that was along the lines of the ideas of the philosopher Alfred Whitehead. Lyttelton continued,

> Three or four of the best brains in the country should do their best to agree on Whitehead's suggestions in the framing of a curriculum and giving their reasons. The outcome would of course be provisional: that is to say the little Committee should meet, say, every three years and compare notes of what they may have gathered as to the desirability of altering in some small respect their original statement of fundamentals or the outline programme based on it.[26]

Such an arrangement, Lyttelton urged, would help to support secondary school teachers who should feel themselves to be "soldiers in an army which is being led by the best generals available, instead of a horde of lost sheep bleating in a howling wilderness."[27]

Lyttelton's proposals met with a characteristic response from the Board of Education, which preferred as usual to avoid direct intervention in curriculum issues. Indeed the idea of the curriculum being prescribed by a small committee was anathema to the chief inspector, F.B. Stead. In Stead's opinion it was far from clear how such a committee should be selected and by whom, and it was unthinkable that its ideas for a curriculum could be imposed on the schools: "Does he think that the teachers would swallow the conclusions of the Committee *sans phrase*? He can hardly mean that this new scheme would be imposed on the schools by the Board? It would be entirely contrary to our traditions and practices to do any such thing."[28] At the same time, such issues were appropriate for the Board of Education's consultative committee, which had been responsible for the major report on education of the adolescent in 1926 and had now completed an inquiry into primary and nursery schools under the chairmanship of Will Spens, the master of Corpus Christi Cambridge. The consultative committee accordingly began in October 1933 to inquire into secondary education, paying particular attention to the framework and content of the education of pupils who did not stay at school beyond the age of 16.[29] It was this that was to become the key forum for debate about the character, ideals, and curriculum of the secondary schools until it produced its major report on the subject, more than five years later, at the end of 1938.

Norwood was not a member of this committee, but he had a significant influence on its ideas and on its final report. This was partly through his direct contact with the committee, although he was only one of 150 witnesses who were called to provide evidence. He sent it a memorandum emphasizing the importance of establishing a definite aim and purpose for the secondary school curriculum, rather than leaving it to be dominated by the requirements of the School Certificate examination. The purpose of the school curriculum, he proposed, should be defined as "training for citizenship," while the school life as a whole should be concerned to develop personality. Moreover, he argued, the foundation of the curriculum should be English culture, in which English studies should be devoted to the training of pupils in speech, writing, and reading. English studies, Norwood added, should include history and geography for all. Only one foreign language should be attempted, but this should be regarded as an essential part of a secondary education. Physical education and hygiene should be given more attention, and all pupils should have training in handicraft, art, and musical appreciation. Indeed, he suggested, these latter subjects were almost as important to him as the English subjects, and more important than the foreign language, or mathematics and science. Finally, he contended, the School Certificate examination should be based on this curriculum, its dominance should be reduced, and it should not be the single means of qualifying for entrance to a university.[30]

These points were emphasized further in Norwood's oral evidence to the committee. He reiterated his ideas about education for citizenship, noting that pupils are the future citizens and so their education should be designed and organized with this end in view. He was also anxious to pursue his argument about the centrality of English culture in the secondary school curriculum and complained that pupils tended to leave school "with only a very slight knowledge of the history and manners of the country in which they live." Moreover, he did not let slip this opportunity to stress the contribution of the grammar schools, and the need to retain a key role for them in the educational system of the future, concluding, "There is a very great moral significance in maintaining the cultural ideal of the better type of Grammar School."[31]

At the same time, Norwood was also able to bring indirect pressure to bear on the Spens Committee, especially through the involvement of the newly established Association for Education in Citizenship. The founder of the AEC, Sir Ernest Simon, was also active in making contact with the Spens Committee, of which his wife, Shena, was a member. His industry was rewarded with a personal commitment from the chairman, Will Spens, expressing his general support for Simon's views. Indeed, Spens went so far as to declare in a private letter, "I entirely agree as to the importance of education for citizenship; if it is not education for citizenship education has no meaning."[32] Armed with this basic agreement, Simon formed a working partnership with Norwood to improve the profile of the idea of education for citizenship, and to develop its practical implications. He agreed to address the third conference of young public school masters held at Harrow in 1935, which was strongly influenced by Norwood's acolytes.[33] Simon also went to great lengths to publicize Norwood's ideas on the secondary school curriculum, with the aim of shaping the outlook of the Spens Committee. At the end of 1936, his wife informed him that Spens thought the curriculum subcommittee, which had been established to look in detail at curriculum issues, "was coming round to the Norwood view." He decided that this was the right moment for a "special effort" to support Norwood's proposals and his general approach.[34]

Simon proposed to do this by the simple device of publishing Norwood's ideas on the secondary school curriculum as a pamphlet under the auspices of the AEC. For this purpose, Norwood provided Simon with a short text, drawn mainly from a speech to the National Union of Teachers two years before, that Simon praised as "a wholly admirable and most challenging statement."[35] As Simon explained to Spencer Leeson, the head of Winchester College, "This is a bold and challenging curriculum and has Dr Norwood's high authority behind it."[36] Nevertheless, Leeson advised caution in associating the AEC with the respected but controversial Norwood. Leeson observed, "Anything that Norwood writes is worth reading and disseminating—but I am not sure whether it would be wise for our Association to commit ourselves to this particular document." In his view some AEC members as well as others who were interested in the matter would take issue with much of the content of Norwood's pamphlet.[37] Despite this, Simon held his ground. He agreed that the association should not take responsibility for Norwood's curriculum but argued that it could still publish the pamphlet "as being by a distinguished educationist and as putting forward views that ought to be considered." In particular, he felt strongly that Norwood's ideas in this form might help to influence some of the members of the Spens Committee, as well as a number of secondary school heads.[38] Leeson conceded the point and acknowledged that "so long as we make it clear that we take no responsibility and are not necessarily committed to agreement with his views, I feel that nothing but good could come of the publication of Norwood's paper."[39] At the same time, his clear reservations were a sign that Norwood's predilection for controversy might undermine support from his colleagues in the public schools as well as in the Board of Education.

Norwood's pamphlet gave a typically forceful account of his criticisms of the secondary school curriculum and went on to outline his proposals for the future. He was insistent that much of the curriculum in the schools was unnecessary for most pupils, especially in the way that it sought to prepare them for more specialized studies, and it failed to teach them what to him mattered the most, namely, the fundamentals of citizenship and their English heritage. The outcome, he alleged, was a curriculum composed of a number of separate subjects, driven by the pressures of examinations, which did nothing to prepare pupils to be citizens in a modern world. The curriculum that he favored was one that would establish the schools as a training ground of citizenship, giving pupils a sense of value of their heritage.[40] English, history, and geography should work together to train for "British citizenship." English would emphasize clearness and directness of speech, history a sense of how the world has come to be as it is today, and geography to find their way around it. In combination, according to Norwood, these subjects would "give to every boy and girl a sense of who we are and where we are and how we have come to the place in which we find ourselves."[41] So far as other subjects were concerned, he went so far as to suggest that foreign languages might be omitted altogether because of a lack of time, that mathematics was taught at too high a standard for most pupils, and that science should encourage citizenship rather than specialization. Overall, the pamphlet summarized the ideas that Norwood had been developing around the curriculum for a number of years, but it is not surprising that Leeson was concerned about the provocative nature of the argument.

The strategy conceived by Simon was successful as a way of bringing these ideas to the attention of the Spens Committee at a crucial stage in the development of its report. The secretary of the committee, R.F. Young, agreed that Norwood's pamphlet was "of first-rate importance to the Committee for their present Inquiry, though of course most of them have a general notion of Dr Norwood's views on the topic in question." Copies of the pamphlet were circulated among all members of the committee.[42] Hugh Lyon, headmaster of Rugby School and a member of the committee, provided welcome confirmation about the potential impact of Norwood's pamphlet, as he told Simon privately, "Much that Norwood says about the importance of the main English subjects is becoming a matter of general agreement and will, I hope, find a reflection in the decisions of the Consultative Committee in due course. I do not go all the way with him for various reasons, but I certainly subscribe to his main principles."[43] Yet it came at a cost in terms of the response from teachers and headteachers who remained unconvinced by Norwood's rhetoric. One such head, A.M. Gibson of Liverpool Collegiate High School, was especially disgusted at the prominence being given to Norwood's ideas. Simon noted that Gibson, "an eminent headmaster in Liverpool and a real friend," liked everything that the AEC had done "except Norwood's pamphlet." Indeed, Simon recorded, according to Gibson, "Norwood is an object of hatred and contempt among most secondary school masters for preaching exaggerated stuff now which he never made the remotest attempt to put into force himself."[44] Once again Norwood was tending to polarize opinion through his vigorous and uncompromising style of debate, attracting committed supporters but by the same token alienating many others.

Despite the privileged nature of his interactions with the Spens Committee, Simon continued to be concerned that its final report might not reflect the points that had been raised about the importance and character of education for citizenship. He even proposed that a further departmental committee should be established by the Board of Education on this specific issue.[45] Spens persuaded him that this would not be possible, much to the relief of officials at the board.[46] In the event, the published report produced by the committee did reflect much of what Simon had been demanding. In its introductory chapter, it pointed out that with democracy facing a challenge, it was the duty of citizenship to ensure that "all should be taught to understand and to think to the best of their ability."[47] Moreover, it continued, it was scarcely possible to exaggerate the importance of education for citizenship in current circumstances, which were such that "all teaching should contribute to this end."[48] It recognized that the foundations for fitting the youth of the country to fulfill their later duties needed to be laid before the age of 16, including the provision of information about national and international issues and local government. This, it proposed, could best be done through established subjects such as history rather than by the general introduction of a civics course.[49]

So far as the details of Norwood's proposals were concerned, there was certainly recognition of his ideas in the Spens report, but also more than a hint of ambivalence. A number of witnesses had no doubt indicated their aversion to Norwood's influence and his general approach, and there was little in the report itself to advertise that Norwood had a key role in its development. In some respects the report seems to have tried to balance rival factions, as, for example, in its general discussion of the need to reconsider the curriculum. It pointed out that one witness had suggested that secondary education had no clear purpose. On the other hand, it conceded,

> we have received convincing evidence that those who are intimately concerned with the conduct of Grammar Schools have clear ideas of the aims which they have in view, and that these aims are concerned with the training of the pupil, first as a person with a body, a mind and a spirit, second as a future citizen of a democratic country, and third as one who will have work of some kind or other to do for a livelihood. With varying emphasis, now on one and now on another of these aspects, this aim is implicit in all schools.[50]

It was also conscious of other influences and developments that needed to be recognized in the further development of the school curriculum, some of which were at odds with Norwood's general philosophy. It noted, for instance, that social and technological changes were taking place much faster than in the Victorian era: "We live at a faster rate, the old lines of social cleavage have become blurred and are breaking down; with improved means of inter-communication and transport the world has contracted and its peoples have been brought nearer together, and their lives, thoughts and actions are in closer contact."[51] Moreover, it also emphasized the advances made in psychology over the past forty years which had given much greater insight, as it seemed to the committee, to the nature of individual differences and aptitudes.[52]

This emphasis on psychology and individual differences reflected the influence of educational psychologists such as Cyril Burt, professor of psychology at University College, London. The report's debt to Burt himself was handsomely acknowledged throughout, and he was given the honor of expounding his ideas on faculty psychology in a ten-page appendix to the report.[53] The secretary of the committee also contributed an appendix on the development of the conception of general liberal education; Professor Isaac Kandel of Columbia University in New York was given the opportunity to expound at length on the secondary school curriculum; H.R. Hamley of the Institute of Education, University of London, developed details of the cognitive aspects of transfer of training; and Fred Clarke, also of the Institute of Education, expatiated on the influences affecting secondary curricula in the Dominions. Norwood was if anything conspicuous by his exclusion from this distinguished company, and he was not even given an entry in the index to the report.

Nevertheless, the report did take account of Norwood's ideas on the curriculum. It explained that the curriculum should be developed around an attitude to life and one main core of learning, with this core being the "English subjects." In its chapter on school subjects that required major changes, it also followed closely the proposals that

Norwood had urged upon it. In the subject of English itself, it opined, "The first aim of all English teaching should be to enable a child to express clearly, in speech or writing, his own thoughts, and to understand the clearly expressed thoughts of others." Moreover, the second aim was "the development of the power thus acquired to benefit the child as a social being, and to help him to take his place as a thinking individual and a wise citizen."[54] It went on to argue that less time should be devoted to mathematics, and that this subject should give greater attention to the utility of mathematical ideas in social life.[55] Meanwhile, it proposed, science should avoid early specialization and should be developed as a broad introduction to the natural laws, as a complement to historical studies, and to impart scientific methods of thought and investigation.[56] All of these recommendations bore a distinct resemblance to Norwood's suggestions. However, the report gave no acknowledgment to Norwood even in this area, and it referred mainly to previous departmental reports to justify its conclusions.

It seems, therefore, that Norwood had become too controversial a figure for his ideas to be drawn on openly in this major report. He retained support for the clarity and conviction of his ideas, but his opponents and detractors were increasingly alienated by them. Although he had risen to the challenge posed by the rise of fascism and the threat of war, he represented nevertheless a previous generation. The middle classes to which he appealed were themselves being superseded by new professional groups more concerned with rapid social and technological changes than with religion or tradition. It was these groups that Spens's report on secondary education was chiefly designed to attract. Norwood appeared to be a waning and marginalized force in the final year before the onset of World War II.

New Ideals in Education

The incipient decline of Norwood's influence in education was reflected in a number of ways in a series of international conferences under the auspices of the New Education Fellowship in which he participated in 1937. This major set of events took him and his wife Catherine to the United States, New Zealand, and Australia during their four months away from home from June until October, and he took the opportunity to rehearse his trenchant and strongly held views upon a wide range of audiences. He was, nonetheless, not quite the star attraction that he had been ten years before, and the long and demanding schedule proved exhausting and a mixed success for Norwood himself. Moreover, Norwood's ideas often sat uncomfortably with the new ideals championed by some of his fellow participants.

The NEF was founded by Mrs. Beatrice Ensor as an international movement, intended to bring together those who believed that contemporary threats to civilization should be approached as problems of human relationships that required new approaches to education in a changing world. It encouraged progressive and child-centered thinking as a key part of its role.[57] Its conferences enabled teachers, other educators, and the informed public in different countries around the world to be actively involved in discussion with acknowledged leaders in the development of educational ideas. These could be significant influences in the further development of education systems. In New Zealand, for example, the events planned for the NEF visit in 1937 served to promote new approaches to the educational policy that was being cultivated by the Labour government, elected in 1935, under the guidance of the minister of education, Peter Fraser, and the educational administrator, C.E. Beeby. Fraser gave strong support to the NEF visit and was encouraged not only by the opportunity that it afforded for the dissemination and exchange of ideas, but also by the enthusiasm shown by teachers and the general public. Moreover, he was aware of the potential benefits of its meetings for the educational reforms that he had in mind, as he made clear to the organizing secretary of the Australian Council for Educational Research:

> The Conference . . . came at a most opportune time, for we had but recently taken our education system under critical review. By no means the least effect of such discussions, so representatively attended and so fully reported, has been the creation of a deep public interest which makes reform possible and subsequent progress assured.[58]

Beeby was also conscious of the intangible effects of the conference held in New Zealand, especially as "it definitely established education as a thing a community can become excited about, a thing that can be discussed in trams and over afternoon tea." As such, he felt, it would do much to raise the status of teaching as a profession, and to "assuage the loneliness of the educational pioneer." Moreover, he reflected, "Quite apart from education in the narrower sense, an isolated country such as ours must have gained something towards international understanding from contact with men and women from overseas so cultivated and so personally pleasant."[59] The Norwoods' participation in this series of events represented a major opportunity for Norwood himself to enhance his standing and authority on the world stage.

The world tour began badly for the Norwoods in the first stage of the journey in the United States. Cyril was robbed of his pocketbook and some money almost as soon as they reached New York.[60] They went on to Chicago, where Catherine felt very ill at ease. She commented disparagingly on the slums of Chicago, and "the dreadful poverty and hopeless outlook of the whole countryside between New York and Chicago, with shacks not fit for humans to dwell in—both here and in various parts right across the continent—a fact which accounts for half America's problems."[61] Cyril's excursion to see the Grand Canyon led him to reflect sourly on a "trip with a lot of vulgar American females."[62] Los Angeles was treated with even greater incomprehension and disdain, as he noted with bemusement, "You can wear anything here and look like anything, no-one seems to notice. Much discussion about capitalism, like Hyde Park, speeches prohibited, but lots of little groups. I never saw so many bad faces, or so many helpless faces in my life before . . . charlatans, cheats, tramps—many with failure written on them."[63] The elderly British tourists were insular and somewhat narrow in their responses to these novel sights and sounds, and generally they did not travel well. They were both relieved to leave for Fiji, where Cyril observed much more contentedly that it was "very pleasant to be back under the British flag again," and Catherine was pleased to visit a Fijian school whose pupils were the "sons of chiefs, splendid specimens."[64] However, Cyril was taken ill on a number of occasions, upset by the heat and the food as well as by news of family health problems back in Britain,[65] and by the time they reached New Zealand he was exhausted and unwell.

The New Zealand leg of the NEF expedition, which to many others was a resounding success, was therefore a particular trial for Norwood. Catherine's diary, which she kept throughout their journey, records her growing concern about her husband's below-par contributions. On a visit to enjoy the countryside of New Zealand, accompanied by Beeby, Norwood was apparently so moody and monosyllabic that Beeby gave up the attempt to talk with him.[66] Norwood then spoke at a lunch, but was inaudible as he kept dropping his head. According to his wife's account of this occasion, "he can be so good that I felt quite sick . . . he ought to be taking first place and is just letting it all slip."[67] Again, a few days later, she was distressed that a seminar that he presented was "marred by his rather 'bored' monotonous tone I have never heard before."[68] To Catherine's relief, he rallied once the party moved on to Australia. At a very high-level function in Sydney, for example, she enthused that "It was a joy to hear him and they would have listened to him for twice as long."[69] In Melbourne, too, he seemed back to normal, as Catherine recorded, "he really is a brilliant speaker and so well now."[70] Yet Cyril himself remained less than impressed by his surroundings, as he complained in Brisbane of a bad-mannered audience and "trivial and unintelligent questions," concluding, "This country needs to get a new set of teachers—too little culture, too much politics."[71]

Norwood's mood may not have been lightened by the differences in approach that were evident not only between himself and his hosts, but also between him and his fellow speakers. The education systems of both New Zealand and Australia were secular and oriented toward examinations, and he found it difficult to convey his religious ideals and suspicion of examinations in this context.[72] He also struggled to make an impact. For example, the *Sydney Mail*, reporting excitedly on the "verbal fireworks" of the visiting speakers, found Norwood if anything rather subdued. His speeches, as it noted, were "spiced with subtle humour and touches of irony," and he was a liberal and even advanced critic of the contemporary social scene, and yet, noting his habitual emphasis on character, physical fitness, and the virtues of the ancient Greeks, it found that "he was doing nothing sensational."[73] Norwood was overshadowed by some of his fellow speakers, who were more in tune with the prevailing ideas and direction of the conferences. In the published proceedings of the NEF visit to New Zealand, for example, it was Professor Kandel of New York who took the top billing with a paper on "School and Society."[74] Kandel argued in this paper that there were two key social functions of education, to conserve and transmit culture and to adapt social conditions to new needs, and that it was necessary now to make a choice as to which of these functions should take priority. In the context of the threat of fascism, he declared, "The issue in a word is between new forms of despotism and tyranny and democracy, whether man shall be enslaved in the interests of reaction or whether he shall retain the hard-won gains in his upward struggle for emancipation and enlightenment."[75] By comparison, Norwood's now familiar strictures on Christianity and the world crisis, and his discourses on physical education, science, and music, were a mere sideshow.[76]

The Crisis in Education

In the late 1930s, too, the social and political climate turned against the English public schools with which Norwood had become so closely associated. This was a sudden reversal of fortune for the schools, which only ten years before had been experiencing growth and optimism. Norwood's own successful publication of *The English Tradition of Education* had been accompanied by the foundation of a new generation of public schools. This included in particular a group of new schools founded by the Rev P.E. Warrington, vicar of Monkton Combe, known as the Allied Schools. Some of these were boys' schools such as Stowe, Canford, Wrekin College, and Seaford College. Others were for girls, including Westonbirt and Harrogate College. Norwood clearly had a strong personal influence over the development of these schools. At Canford, for example, the strong Anglican character of the school was challenged when the headmaster appointed a member of staff to the position of senior housemaster and it was then found that the new staff member had been a master at a Roman Catholic seminary. Norwood wrote to the council of the fledgling school to recommend that the headmaster should resign, and this advice was accepted.[77] The following month, Norwood wrote again to support the appointment of his own son-in-law, the Rev Clifford Canning, as the new headmaster, and this was also agreed to.[78] These remarkable developments reflected Norwood's personal stature as well as a clear element of patronage and even nepotism in the management of such schools. Norwood's personal connection with Canford School was strengthened further in 1932 when the youngest of his three daughters, Barbara, married Alexander Barbour-Simpson, an assistant master at the school. Norwood also took a major role in the Allied Schools network as a whole. In 1934, prompted by financial difficulties, the schools came under centralized management, and Norwood was invited to become the chairman of the central committee.[79] He held this position for the next twenty years.

Norwood also continued his efforts to bring the public schools closer together with other secondary schools, although by now he was acutely aware of the resistance to such a mutual accommodation. As he noted in 1935, anyone who attempted to deal with this problem "lays himself open to misunderstanding in several quarters, and must not mind rebuke or abuse."[80] Nevertheless, he argued that his own unique professional experience of sixteen years teaching in secondary schools and a further seventeen years in public schools entitled him to make such an attempt, and, he added, "I mean to be bold enough to set down what I think, even though I stir up opponents in both camps."[81] Norwood pointed out that since the Education Act of 1902, there had developed

> two types of education utterly cut off from each other, yet doing the same work—one leading through the elementary and secondary school, and the other through the private preparatory school and the public boarding-school, to the university, where the products tend to meet like strange dogs, or to the professions and to business, where they are at any rate in danger of forming separate castes.[82]

If anything, he warned, the system had hardened in recent years, producing a "deepening fissure" that would prevent the establishment of a single purpose and a united nation.[83]

Thus, although Norwood no longer had direct involvement as a headmaster, his years of experience and his continuing close connections with the public schools led him to take an active interest in the future of the public schools. When they encountered difficulties, therefore, Norwood was quick to register alarm. By the late 1930s, enrolments at the schools began to show signs of decline, caused by a combination of low birth rates and financial difficulties. The schools also came under increasing criticism for being out of date and elitist. As the international crisis gave way to open warfare, they were widely viewed as being responsible for Britain's difficulties. Norwood responded by calling for further public schools reforms that would bring them under the control of the State. Yet he found himself distrusted by both the defenders of the public schools and their critics, and vulnerable to attack from both sides.

In addressing the problems of the public schools, Norwood tried to enlist the support of his contacts at the Board of Education. At the end of 1938, he wrote to a senior inspector, F.R.G. Duckworth, to ask for a Royal Commission to be set up to inquire into the public schools. He emphasized his concern at the decline in the numbers of enrolments and the resulting lowering of standards: "Headmasters are literally spending half their time in commercial travelling and touting on Prep School doorsteps."[84] Invited to the board to explain his ideas, he took the opportunity to elaborate on the problems faced by the schools.[85] His intervention had an immediate effect, as Board of Education officials recommended setting up a commission to investigate these difficulties. However, Norwood was ruled out as a potential chairman of such an enquiry. As the officials observed, "Sir Cyril himself is an obvious possibility, but he is intimately connected with one group of schools [the Allied Schools] and he is by no means popular in some quarters."[86] In the end, these plans came to nothing, and Norwood had to come to terms once again with the decline in his own influence in the educational world.

War was eventually declared in September 1939. Despite the immediate threat of German invasion and possible defeat, the war produced a renewed interest in the need for radical reform and the future development of education in a postwar society. For example, Fred Clarke, the director of the Institute of Education at the University of London, published a short book on education and social change at this time, and engendered a wide and positive reponse.[87] Norwood also adopted a view that put established structures such as the public schools into question, as he again set about eliciting support from the Board of Education. He had always feared the collapse of the ideals that he had championed. He now went so far as to express doubt that the public schools had a future in their current form: "I do not see how as a system they are to survive this war. They have had their hundred years, and have rendered very great service with certain limitations. But the order of society to which they belong is at an end, or near it, however we may regret the fact."[88] He insisted that a system based on scholarships would be ineffective, "like a speculator who is trying to hold up shares in a market out of which the bottom is falling," and would be avoided by the leading schools. Norwood recommended instead that a national system of boarding schools should be introduced under the auspices of the Board of Education, for boys from the elementary schools, and the sons of clergy and ministers and other social classes for a small fee. This, he suggested, should not be administered by the LEAs, "who would not understand it." The aim would be to democratize the system without destroying its values, as part of a common education system for all after the end of the war: "I don't know whether we are going to win the war, or in what sense we shall 'win,' but I do feel that we are not wise in trying to stop temporary leaks when we have a whole ocean coming in." Moreover, Norwood advised, the Board of Education would need to take an active lead in these major developments. Indeed, according to Norwood, "We shall need a statesman of real vision at the Board to lead us, when the time comes."[89]

In the early months of 1940, Norwood risked further rebuff as he pursued these ideas in the public domain. In a two-part contribution to the weekly periodical *The Spectator*, under the title "The Crisis in Education," he called for a Royal Commission to be set up to consider the whole field of education. In particular, he proposed that this should address the conflict between the public schools and State-aided secondary schools, which had fostered social class divisions and inequality of educational opportunities. They should now be grafted together to provide "that single national system of the future which will make democracy not unequal to its task."[90] The second part of his essay also clarified details of the education system that he envisaged would emerge after the war. Formal education would continue up to eighteen, followed by national service. The first stage of education would be for pupils up to or just before the age of twelve, either in public elementary school or preparatory school. This would be followed either by six years of secondary education, or by a four-year course leading to two years of part-time continued education. The public schools would become a "vital part" of the system by changing their age of admission and organizing a four- or six-year course on the same basis as the secondary school, accepting up to 10 percent of pupils from the elementary schools and giving financial responsibility to the Board of Education. Finally, he concluded characteristically, such a system should be inspired by the ideals of public service and ultimately by spiritual values.[91]

These were far-reaching and radical proposals and predictably met with the displeasure of the more entrenched supporters of the public schools. One such supporter, Lord Hugh Cecil, the provost of Eton College, was moved to protest to the president of the Board of Education, Lord de la Warr. Cecil vigorously denounced the notion of a Royal Commission, or any State policy at all that would impinge on the public schools. He proceeded to castigate Norwood's articles in *The Spectator* as "pure Totalitarianism." According to Cecil, the education system should be based not on unity but on variety. He argued indeed that not only should the public schools continue to operate on their existing lines, but that all secondary schools should be set free from the educational control of the State. As for Norwood himself, Cecil was dismissive: "It is not surprising that Sir Cyril Norwood's scheme culminated in compulsory Military Service. The Brown House at Munich is evidently his spiritual home."[92] Lord de la Warr went out of his way to be reassuring in his response to the affronted provost, adding for good measure that Norwood did not have official support and did not represent any wide body of opinion: "I don't think we need attach too much importance to some of the matter which has appeared in the press, and, so far as Sir Cyril Norwood's articles are concerned, I gather that he speaks for no one but himself."[93]

This interchange was a further indication of the weakness of Norwood's position in terms of the suspicion in which he had come to be held in the elite circles of educational policy. His combative and uncompromising approach to debate, and his habit of speaking his mind in the public media, had always disconcerted officials, politicians, and public schools that much preferred to discuss educational problems in private. These tendencies had finally alienated many who he needed to give sympathy and support, or at least acquiescence to his views. It would, however, be untrue to suggest that his ideas were completely isolated. The *Times Educational Supplement*, for instance, also voiced criticisms of the public schools for the segregation that they imposed and the unequal opportunities that they represented in training for positions of responsibility and leadership.[94]

Yet these criticisms of the public schools also represented a further aspect of Norwood's exposed position, for the critics of the schools were just as suspicious of his ideas as were the diehard adherents. One fierce opponent was T.C. Worsley, formerly a pupil of Norwood at Marlborough College. Worsley declared that the public schools were largely responsible for the failures of leadership that had led to a second world war, and that they were both undemocratic and irrelevant to a modern society. This critique was also damaging to Norwood's approach, for Norwood had stoutly defended the continued existence of the schools and had seen in them the prime basis of the English tradition of education. Indeed, Worsley went out of his way to point out the failings of this tradition and to undermine Norwood himself for his attachment to it. According to Worsley, the ideas put forward in Norwood's major book summed up "the spirit by means of which he became, in the first post-war decade, the leading educational figure."[95] Yet, with the growing economic crisis and international conflict of the 1930s, Worsley continued, even while Norwood was writing his book, "the first whiff of the coming disaster must have been blown in through his study window."[96] Moreover, Worsley suggested, the Platonic implications of the ideal of leadership represented in the public schools were themselves inherently suspect. He argued that that the elitist connotations of the schools not only clashed with democratic ideals but were similar to those of the Nazi regime in Germany. Thus, he contended, "The latest example of this kind of planning has been provided by the Nazi movement. A year ago Hitler created his special leadership colleges on Platonic and Public School lines."[97]

Norwood responded to Worsley's open attack in a spirited fashion. In a review of Worsley's book in the *Journal of Education*, Norwood subjected it to detailed criticism. Worsley, he noted, "wields so vigorous an axe that he leaves nothing standing." In Worsley's book, 126 pages were devoted to a "slap-dash" history of the public schools, 82 to the debunking of their present pattern, and a mere 26 to the plotting out of the future of education, with the final 46 pages being "a sort of wheelbarrow in which the author carts away the public school rubbish to its appropriate tip." Overall, Norwood averred, "The matter shows all the faults of hot youth: on almost every page there are instances of exaggeration and hasty misrepresentation."[98] He was also able to point out that he had been criticized by supporters of "Tory orthodoxy" in the field of the public schools for the past twenty years, and so he added, "I find it refreshing to be exposed as after all the chief reactionary."[99] However, this defense of his own work only served to underline his vulnerability. He had been deserted by the influential officials of the Board of Education and isolated by the leading public schools, and now he was cast as the principal villain by the critics of the schools.

Furthermore, the terms in which Norwood now found himself denounced from all sides were themselves significant. Norwood had been the chief celebrant of a tradition that he defined as uniquely and characteristically English in nature. He had also instinctively opposed the new fascist regimes from their inception and called for an alternative approach to education for citizenship that would counter their ideals in a democratic state. It was somewhat ironic, therefore, that he was now criticized for being both un-English in his approach and sympathetic to fascist ideas. Lord Hugh Cecil saw the Brown House in Munich as Norwood's spiritual home because of his emphasis on the role of the State; Worsley drew connections with Hitler's initiatives in Germany because of the elitist implications of Norwood's approach. The underlying reality, as the debate over the future of education became increasingly polarized and heated, was that Norwood's attempts to find a basis for unity became unsatisfactory for both sides. Despite his best efforts to respond to the new challenges of the time, Norwood increasingly seemed marooned and out of place in the new world of education.

Chapter 9

The Norwood Report and Secondary Education

Norwood's influence was clearly waning both in elite policy circles and among a newer generation of opinion formers. In the relative backwater of his old Oxford College, he faced retirement and a quiet end to his educational career. Yet the war gave him an unexpected opportunity to define the future of secondary education. He was invited to chair a committee that became a key forum for the major reforms developed in response to the new radical ideas that were circulating about education. His contribution to this wartime debate gave renewed force to the ideal of secondary education that he had cultivated in changing circumstances over the past forty years. It also gave rise to hostility that challenged and eventually undermined the reforms that were established under the Education Act of 1944.

Oxford Men

As Norwood had calculated when he made the decision to leave Harrow and become the president of St. John's College Oxford, he would be taking up "a position of comfort and dignity with abundant leisure."[1] This proved to be the case, and he was handsomely rewarded in material terms. His salary when he was appointed was £ 1900 per annum, increased further to £ 2000 two years later.[2] He also had full use of the president's lodgings (rent and rates free) that had been recently renovated with the installation of central heating.[3] This was itself a considerable asset. One admiring visitor, F.B. Malim, the master of Wellington College, described it as "almost the most lovely house in Oxford, with wonderful portraits."[4] At the same time, Norwood was provided with a butler and personal servant, and he was also able to employ a chauffeur and general handyman. This was substantial comfort, and Norwood had the additional security of knowing that he would be entitled to hold office until the age of 70, with the option of reelection for a further five years by mutual agreement.[5]

His post was also one of high status, within the university and beyond. St. John's itself was among the richest colleges in the university and owned a large amount of land in North Oxford.[6]

These were compensations for a daily routine of committee work within his college and for the university as a whole. He was soon elected, for example, to the university's committee for rural economy, and then as the university's delegate of home students.[7] From 1937, he was a full member of the university hebdomadal council.[8] He became a member of the committee on physical exercise, the committee on the requirements of residence, the committee on city questions, and the committee for ornithology. In 1939, he was appointed as the chairman of the Oxford Society, which provided a point of connection between the university and the colleges.[9] The following year, he was appointed as the council's representative on the Oxford advisory committee to Makerere College in East Africa.[10] By 1946, when he retired, Norwood was on three out of thirteen standing committees and six of the twenty other committees, or more than a quarter of the total. During this time, naturally enough, he also played a part in the development of the university education department. The title of the department was changed from the Department for the Training of Teachers to the Department of Education, in order for it to play a broader role in the study of the aims and principles of education, the conditions of education, the history of education, and both practical and theoretical issues relating the methods of teaching, organization, and administration.[11] In February 1939, Norwood himself was elected to be the director of the Department of Education and was given the title of reader in the Department of Education at the university.[12]

Besides these mainly administrative activities, Norwood also found time to promote broader interests. For example, his advocacy for physical education, previously centered on the schools, was now transferred to the university. He pointed out the importance of physical education, which he described both as a new conception of education and as Platonic in inspiration, and argued in favor of developing a central university institution for advice, guidance, and training in this area. In particular, he called for the establishment of a gymnasium and a swimming bath to serve students across the university. This he also saw as being vital for the training of future teachers in physical education at the university's Department of Education: "You might as well teach Physics without a laboratory."[13] There was, however, some resistance to this idea, not least from another member of the university, Nevill Coghill, who claimed that physical exercise was a "dangerous drug" and was "almost fatal to the mind in large quantities."[14] Another consuming interest, religion, Norwood pursued from 1937 as the president of the Modern Churchmen's Union. In this capacity, he worked out his Christian beliefs as a "comprehensive, central and critical Churchman," that is, as a Christian Modernist, in a spirit that was regarded by sympathizers as spiritual, moral, and rational.[15]

Norwood developed these active and worthy pursuits despite being increasingly reserved in his personal relationships. One acquaintance, the prospective head of the geography department at Canford School, was invited to be a guest at St. John's College for more than a week, but he failed to make an impact on the taciturn

Norwood. As he later recalled,

> I saw Sir Cyril at breakfast where I appeared at 8 o'clock—he at 8.05 when he grunted a short welcome, thrust his face behind The London Times and set about a fairly large plate of porridge! I tried conversation on the affairs of the day but he was reluctant to join in and so I gave up any real contact at about eight or so consecutive breakfasts![16]

This largely humdrum and quiet lifestyle was disturbed from 1939 onward by the onset of World War II. The controllers of fish and potatoes at the Ministry of Food were billeted in several buildings of St. John's College, and a nightly blackout and a much more Spartan existence became routine.[17] The number of undergraduate students declined, and those who remained often became restive at the stuffy college atmosphere of wartime Oxford. College authorities such as Norwood and the other mainly elderly dons who continued to enforce college restrictions found themselves the targets of mischief and abuse.

At St. John's, among the most prominent of the student dissidents at this time were the future poet Philip Larkin and the future novelist Kingsley Amis, who found a common cause in deriding the values and rules of their elders. Both had contempt for what Andrew Motion, Larkin's biographer, describes as "the snobbish, public school elements of Oxford life."[18] The suggestion books of the college's senior common room provide ample testimony to the mockery to which Norwood and other college fellows were subjected. Even before Larkin's arrival in 1940, the following verse was inscribed in the book:

> If accosted by Norwood
> Any self-respecting hor [sic: whore] would
> Charge double
> For her trouble.[19]

In 1941, much disrespect was shown after a mawkish poem by Norwood was published in *The Times* under the title "To a Fallen Airman." The poem read,

> High o'er the clouds, and aflame with the fire of sunrise
> Poised on the edge of the moment, to die or to live,
> Sheer with the Peregrine's stoop through the swirl of the skies,
> Daring the cost, you struck home: what more could you give?
> Dead: but the flame of your youth cannot fade with the years,
> Dead—but more living in spirit than those who draw breath,
> Perfect by sacrifice, cleansed of our doubts and our fears,
> Fallen, the ransom of many, and Victory o'er Death.

This was immediately pasted into the suggestion book with a message: "Sir, if The Times persists in printing such utter balls as this I suggest it should not be purchased in future." Larkin then gave notice of his mordant humor with a sarcastic riposte to this message: "Sir, I must protest against Mr Hughes' blurted condemnation of the above moving piece of verse, and his implied insult to a figure we all respect and honour.

Surely, Sir, the ability to scan throughout eight lines is not to be despised, and to produce three good rhymes out of four is no mean achievement." Moreover, Larkin continued, "Here is a sensitive mind exceptionally and beautifully alive to the splendour and nobility of youth, competent and eager to immortalise in verse his most moving prescriptions; all must appreciate, in sum, this brief yet profoundly spiritual expression so native to our own times, yet surely in feeling so akin to eternity." Nevertheless, Larkin's own true feelings were highlighted in his gleeful postscript: "The dirty old man."[20]

These clues to Norwood's life and position in Oxford at this time are significant for the evidence they provide of his changing profile in relation to the outside world. He was now unrecognizable as the ardent reformer that he had been so many years before at Bristol Grammar School and was identified simply as one of the "old autocrats."[21] He was to all appearances the epitome of the established order, and as such a prime target for irreverent spirits and radical critics of society. Moreover, his growing remoteness and aloofness, apparent since at least the 1920s, made him seem out of touch, and careless of the concerns of the rising generation. These were outward indications of a hardening conservatism in his approach to education, as the fresh ideals of forty years before became familiar, rigid, and increasingly old-fashioned. Retirement now beckoned him, certainly into comfortable and dignified circumstances, but somewhat removed from the influence and attention he had previously enjoyed. Yet there was to be a sting in the tail, as Norwood was called back to intervene at a crucial stage in the making of new and major reforms of the education system.

The Elder Statesman of Education?

The Spens Report had already mapped out the problems of secondary education and made a number of recommendations for its further development. It came to the conclusion that the grammar school curriculum should be reformed so as to correspond more closely to changing social and economic needs, and that the School Certificate examination would also require radical changes in order for this to take place. Modern schools would be an alternative, nonacademic form of secondary education, as the Hadow Report of 1926 had already proposed. At the same time, it recommended that junior technical schools should be elevated to the status of secondary education as technical high schools with equal status to the grammar schools. It insisted, moreover, that all secondary schools, although of different types, should operate under the same conditions and enjoy what it termed "parity of esteem."[22] Indeed, it suggested, "The barriers between different types of secondary school which we seek to remove are the legacies of an age which had a different educational and social outlook from our own."[23] It concentrated therefore on resolving a range of "administrative problems" that might undermine the parity of different kinds of secondary schools in a reformed system.[24]

Nevertheless, there were many who doubted that it would be possible to achieve parity of esteem between schools with different approaches and outcomes. At least one member of the Spens Committee itself, Lady Simon, harbored deep reservations about the likelihood of such parity being brought about.[25] A number of senior board officials were also doubtful. One, Robert Wood, was worried that such a system might be perceived as "nothing more than an elaborate attempt to spoof the public mind, and by giving a new wrapper and a new name persuade them to accept the old cheaper article as the higher class goods."[26] Anxieties were also voiced that the grammar schools might be undermined. For example, another board official, G.G. Williams, warned against going too far in seeking to correct the academic bias of the secondary schools:

> The boast of our traditional Secondary Education was that it taught boys and girls to think and reason for themselves, to distinguish between false and true, to know something of the world about them, and to grow up as well balanced individuals ready to take their place as men and women and as citizens in the community. For many pupils this type of education may not be most appropriate; but to say that we are justified in excluding it in favour of a strongly vocational type of school for all pupils is to lose one of our most precious heritages.[27]

The logic of the Spens proposals to support the development of technical high schools was also open to criticism, as it was not altogether clear why this would be necessary if the curriculum of the existing secondary schools was to be overhauled to meet changing social and economic requirements.[28] As educational reform assumed increasing urgency following the outbreak of World War II, the problematic nature of the Spens proposals came to be seen as a potential obstacle to change.

Norwood was no longer as close to the policymaking elite as he had previously been, but he remained, after almost twenty years, the chairman of the SSEC. It was this position that enabled him to become actively involved in the wartime discussions around reform. At the end of 1940, it was agreed that further consideration was needed about future policy on secondary school examinations, notably in relation to the established method of conducting these examinations through eight separate examining bodies. In any case, it was proposed,

> it is difficult to resist the conclusion that at this stage, when national educational problems are being generally re-examined and overhauled, some formal review of the system of School Examinations after nearly a quarter of a century is called for, and that the Chairman should be empowered to approach the President of the Board with the request that appropriate steps for this purpose should be taken.[29]

Norwood took up this suggestion and was rewarded with an invitation from the president of the board to chair a small committee on school examinations.[30]

From this relatively small beginning, further developments during 1941 brought Norwood back to the center of attention. First, he made a fresh excursion into the public media to explore general aspects of educational reconstruction. A lengthy article by Norwood in *The Fortnightly* published in February 1941 included little in the way of new ideas, but it was a useful reminder of the broad scope of his thinking. It was also noticeably less combative in its style than many of his contributions to the printed media, as if attempting to strike a more conciliatory note that might be acceptable to critics of a broader range of views.[31] He would have been encouraged by the favorable response of sympathizers such as Dorothy Brock, the headmistress of Mary Datchelors School in London and one of his few female intimates. Brock was quick to point out that this was an opportunity for Norwood to provide experienced counsel for the phase of reform that was about to begin. She told him rather effusively, "One thing is clear—that whether you're 66 or 666 this is a time when you are needed.... No one has your wisdom and clear judgment—and no one has your experience. Don't 'get out' just yet—so many of us *trust* you."[32] Norwood embodied experience and continuity, but he could also articulate a vision for the future, and this combination might help to give clear resonance to educational reform. No doubt there were some officials at the Board of Education who also recognized that his conservative tendencies might help to offset the radicalism of some proposals for a new education system that were beginning to be circulated.

A key step toward Norwood's increased participation was the appointment in July 1941 of R.A. Butler as the new president of the Board of Education. Butler was determined to take forward major reforms in the education system, although the prime minister, Winston Churchill, was initially hesitant.[33] Butler was also a former pupil of Marlborough College during Norwood's period as master of Marlborough. Butler seems to have seen Norwood as a potential elder statesman to guide the reforms, and Norwood was happy to oblige. Thus it was that within two weeks of his appointment he discussed with Norwood the idea of broadening the remit of his committee on school examinations, to include "some mention of the content of education as taught in these schools."[34] Norwood in turn reflected on the need to "make the dry bones live," adding, "I suppose Butler is feeling after that, for if you are to ask Parliament for a hundred millions or more after the war, you must make the young men dream dreams." He himself, the "old man," would therefore "try to see a vision" and expand the role of his new committee as a means of bringing this about.[35]

It was soon agreed that Norwood's committee would advise on changes to the secondary school curriculum as well as to examinations, so that it would consider "suggested changes in the Secondary School curriculum and the question of School Examinations in relation thereto."[36] Such a remit, as Norwood explained it to Butler, would avoid giving the impression that the committee would repeat the exercise that Spens had recently completed and would instead enable it to "get our minds on the whole problem."[37] For his part, Butler emphasized that he wished the committee to have the benefit of Norwood's long experience of education, and that he would not limit the scope of the report that it would produce.[38] However, he proceeded to restrict it by resisting Norwood's suggestion that the committee could also refer to the public schools by commenting on the need for a boarding system in British education as a whole. As Butler himself recorded this conversation,

> I said that I liked this approach very much but that I did not want Dr Norwood to produce a solution of the Public Schools question. Neither the Government nor the country were ready for this. Moreover, the solution would be bound to be many handed. All I desired was an approach made. Dr Norwood said that this was what he had in mind.[39]

At the same time, Norwood proposed that while his report would not study the technical and modern schools in detail, it would, however, indicate that all education for pupils over the age of 11 should be regarded as secondary, with a uniform curriculum for all those between the ages of 11 and 13, "so that all children at a later age could feel that they had passed through the same mill at one period in their lives."[40]

The Making of the Norwood Report

Norwood's committee consisted of 12 members including himself. It included the head of a boys' secondary school, Terry Thomas of Leeds Grammar School, and the head of a girls' secondary school, M.G. Clarke, from the Manchester High School for Girls. Miss O.M. Hastings, A.W.S. Hutchings, and E.W. Naisbitt served on behalf of assistant teachers. P.D. Innes (chief education officer for Birmingham), Joseph Jones (chairman of the Brecon education committee), and Percival Sharp (secretary of the Association of Education Committees) represented the interests of the local education authorities. Finally, examination bodies were included in the form of J.E. Myers of the Northern Universities Joint Matriculation Board, S.H. Sharrock, secretary of the Matriculation and School Examinations Council at the University of London, and W. Nalder Williams, secretary of the Local Examinations Syndicate at the University of Cambridge. This constituted a group with considerable experience of secondary education, if not of wider interests, but there was no doubt that Norwood himself was the dominant force on the committee.

It was evident from a very early stage that Norwood had an expansive view of his committee's brief. At the first meeting, while acknowledging that the territory of the Spens Report could not be covered again, he proposed that the committee "had an opportunity of reviewing the whole field of secondary education after the War." He also set out the issues with which it would be concerned, which themselves rested on assumptions long associated with Norwood. Included in these were questions of how to fit independent schools into a unified system, and whether it would be possible to design a general examination for all within a diversified system of schools. He was also emphatic in defending his own most treasured ideals. For example, he declared, "Freedom was essential; yet the tradition of secondary education, the best in the world, must be preserved." Within this basic set of priorities, he proposed that the committee should define a path from primary education to various forms of secondary education.[41]

In setting out the agenda for his committee, Norwood developed some of his most detailed discussions of the distinctive characteristics of the English grammar school. The overall pattern was marked out by G.G. Williams, the principal assistant secretary in the secondary branch at the Board of Education. Williams pointed out the variety of institutions and financial control within the current system, and the difficulties involved in resolving all of its many anomalies. So far as the general lay out of the system was concerned, Williams proposed a federal system, supervised and coordinated by the Board of Education, that would include grammar schools and also public and other boarding schools.[42] Norwood supported this approach, arguing that the committee should aim at "substituting a federal system for the present uncoordinated medley of schools." Under a new arrangement, he added, "each type of school would be enabled to make its fullest possible contribution to national education; and a general standard of efficiency could be gradually secured without destroying the variety and independence which are characteristic of the English system."[43]

Further to this overall framework, Norwood sought to establish "first principles" in terms of the underlying purposes of education "to help each individual to realise the full potential of his personality," through membership of a society and through work in a society. This pursuit of basic principles also led him to emphasize truth, goodness, and beauty as absolute values "based on a religious interpretation of the world which for us must be the Christian religion," and thus to challenge the alternative theory that he summed up as holding

> that there are no absolute values; that truth, goodness and beauty, are all relative; that education has hitherto been backward-looking, designed to confirm and entrench established names; that the new world requires a forward-looking education, based on science, using scientific method to adapt conduct to changing needs, and knowledge to emerging circumstances, . . . that biology and psychology should replace Christian teaching.[44]

It was this latter theory that Norwood regarded as underpinning the Spens Report. Indeed, his advocacy of absolute and timeless religious values steered his committee toward a direct conflict with the secular and scientific approach represented by Spens.

Norwood also attempted to spell out for the benefit of his committee the kind of pupil who would benefit most from a grammar school course, and what such a course should include. Generally, he claimed, the kind of treatment given to studies in a grammar school was one that dealt largely in abstractions. From this, he, therefore, deduced that "it would seem to follow that the Grammar School is best suited to the boy or girl who shows promise of ability to deal with abstract notions—who is quick at seizing the relatedness of related things." Here again he openly contested the view of the Spens Report that the established grammar school curriculum was out of date and lacking in relevance. To Norwood, the point was that to grasp the subject as a whole it might be necessary to spend time on aspects of the subject that might not arise again in everyday life:

> Not much grasp may be necessary to apply a rule or formula or principle to a usual set of circumstances, and it may be important that most should be able to do it; on the other hand, a grasp of the rule or formula in such a way as to enable a person to apply it with necessary adaptation to a set of circumstances presenting some unusual feature may come only from a study of the subject as a coherent and systematic whole, that is, from a study which includes those parts often classed as academic.[45]

Such a treatment, which he defined as "grammatical," would also be more concerned with knowledge for its own sake than any other treatment, and he acknowledged that this would also lead it being labeled as "academic." He concluded that the numbers of pupils in grammar schools and the number of subjects dealt with in this grammatical way should be reduced, while for the nongrammar school pupil, there should be further thought about a "new treatment of a new curriculum suited to his needs."[46]

These early signs of the role being taken by Norwood's committee led to opposition and hostility from rivals. The first of whom was Will Spens, who soon became alarmed at the prospect of Norwood usurping his own recent contribution to secondary education. M.G. Holmes, the permanent secretary at the board, in a note headed "Spens v. Norwood," acknowledged the "rivalry" that existed between the two men, although he added that "any animus that there may be is on Sir Will's side." This had arisen, according to Holmes, because "As Chairman of the Board's Consultative Committee he feels that on any question of importance the Board should consult him rather than anyone else, and he is probably annoyed because the terms of reference of Sir Cyril Norwood's Committee have been enlarged by the inclusion of the curriculum of Secondary Schools, a subject which he regards as peculiarly his own." Nevertheless, in Holmes's view, "I feel pretty clear that the educational world would be more ready to put their money on Sir Cyril Norwood than on Sir Will Spens."[47] Williams agreed that "there is no reason to think that we are acting improperly vis-à-vis the Spens Report."[48] Spens's resentment and suspicion did not abate, and eventually he approached Butler himself to protest at Norwood's activities, but to no avail.[49] Beyond the personal rivalries, the key issue involved in this hostility was the kind of middle class that the secondary school was to represent. Norwood's ideal reflected the religious and provincial, Spens the more secular, scientific, and metropolitan.[50] The future lay with Spens, but for the moment at least Norwood was back in favor.

Despite these differences, there was a general assumption that the proposals of the Spens Committee for three different types of secondary schools—grammar, technical, and modern—would be carried out. However, the further development of radical ideas about education during the war meant that there was growing interest in the notion of multilateral schools that would combine these types of approach. Norwood was strongly opposed to such a development and insisted that multilateral schools had not succeeded in other countries, including the United States. Indeed, he took satisfaction from the skepticism of his colleagues concerning the merits of multilateral schools. For example, he noted that Percival Sharp, the secretary of the Association of Education Committees and thus with a key role in the approach taken by LEAs, was opposed to multilateral schools. As he happily explained to Butler, Sharp was "all against multi-lateralism," because he had observed it in New York and elsewhere and did not believe in the "multilateral headmaster."[51] Even more emphatic was Norwood's account of Sharp's views provided for the benefit of Williams: "He spits on multilateralism, having been to America."[52] Norwood's distaste for the American tradition and style of secondary education was clearer than ever, and this distaste was evident in the advice he gave on the reconstruction of secondary education in Britain at this key stage in its development.

Another idea that had arisen at this time was that of common schools, which would combine the primary schools and the preparatory schools, as a basis for pupils then going into the different types of secondary school. Butler was attracted to this proposition, but he hesitated as he feared that this would not win parliamentary approval, and also that "the law would be got round and people would start up a series of private schools, less satisfactory than those existing at present."[53] This issue went to the heart of educational debates about how far to try to induce social change and equality and how far to preserve freedom and variety. Butler, a forward-looking Conservative, was conscious of trying to balance these aims. He reflected, on the one hand, "in framing educational policy I am continually inspired by relics of the belief that we are a free country," and, on the other, "I do not think that we should underestimate the extent of the social revolution through which England is passing. Moreover we have to consider the position of England vis-à-vis England and the Dominions." Overall, he concluded, it was important to find a way to support freedom, change, and equality all at the same time: "I feel convinced that any educational measure will have to make it clear that we in the Education Department are not afraid of change; that we wish to preserve quality; that we have not in this country two absolutely separate types of education; and that they converge at various points."⁵⁴ His permanent secretary, Holmes, was also equivocal, observing that demands for social equality were likely to grow the longer that the war continued.⁵⁵ Norwood's contribution to this debate was to propose that it would take twenty years to introduce common schools of this type in a satisfactory way, and he appears to have persuaded Butler that such reforms should be delayed so that in due course they would "follow upon an improvement in the schools into which we desired to introduce such changes."⁵⁶

With the support of Butler and officials at the board, Norwood was able to ride out criticisms of his own role. Williams was particularly impressed with Norwood's approach to the task with which he had been entrusted. He was clearly aware of Norwood's dangerous penchant for making independent and outspoken public remarks, but he was reassured that Norwood was not likely to break ranks in the current delicate situation: "Like other eminent public men CN enjoys striking matches to dazzle the eyes of moderate folk, but he is shrewd enough not to burn his own or any other responsible fingers." Williams advised Butler and Holmes that "fortunately" Norwood realized the importance of dividing the eventual report into sections that could be treated separately, in case some involved policy proposals that might be out of line with the educational reforms as a whole, and that in general "we can trust the good sense of Norwood."⁵⁷ Despite these supportive and even affectionate remarks, there is still a sense that Norwood was being kept on a tight leash. Butler's protection was decisive. At an early stage he comforted Norwood that most committees had to overcome outside criticisms, and he emphasized that, in spite of Spens's objections, Norwood's report would be published to provide "practical advice" for the reforms.⁵⁸

A further point of dissension was over the future of examinations, which involved sensitive issues around the role of teachers and the established examination boards. Norwood was determined to abolish the School Certificate examination and to encourage teachers to play a greater part in assessing the educational progress of pupils. This was in line with his views that were often vociferously voiced during the past twenty years or more. At he same time, he was uncomfortably aware that not only would the examination boards strongly resist such a move, but that the teachers would also be difficult to persuade. He was especially impatient with the teachers' conservatism that went against what he argued was an essential ingredient of the professionalism that their further development required. To Butler, he made a "direct onslaught" on the teachers, and an impassioned plea to reform the teaching profession. He was uncertain whether he would be able to secure agreement about examinations even within his own committee but was resolute in his desire to alter the existing system.⁵⁹ A separate committee on the training of teachers and youth leaders was established in 1942, and Norwood wasted no time in giving it the benefit of his opinions about the teaching profession and the role that it should take in matters of assessment:

> Sir Cyril did not believe that the status of the teaching profession could be raised until the teachers were given more responsibility and made to face up to it. At present teachers taught to an Examination standard which was set by an outside Examining Body. Teachers should themselves be able to work out a scheme of education and to assess it. If teachers could be trained to do this, they might in time become a self-governing profession and be esteemed as such; but they were not ready for that yet.⁶⁰

As an interim measure, he proposed that regional councils might supervise teacher assessments, with the School Certificate examination being phased out over five to ten years: "The advancement of such a state of affairs would take time, but it was the only certain way of raising the status of the teaching profession and the dignity and self-respect of the teachers."[61] On this set of issues, Norwood was an advanced and radical thinker. It was common to emphasize the importance of the schools rather than the State in the process of determining the curriculum. Norwood took this a stage further to recognize that in order for this to be meaningful, the strength of the examination system must be greatly reduced and teachers should take a fuller role in pupils' assessment.

A lengthy meeting of Norwood's committee at the beginning of September 1942, in the palatial comfort of the president's lodgings at St. John's, confronted the issues around the future of examinations and also the criticisms that the secondary school curriculum was facing. It agreed that some of the criticisms included in the Spens Report and put forward by witnesses to the committee were "untrue or exaggerated," although some had substance. It also expressed general agreement to some basic principles about the school curriculum, in particular to the idea that there should be greater freedom for schools in building curriculum. It accepted the view, highly characteristic of Norwood's own approach, that there should be three essential elements of education that took precedence over individual subjects: training of the body, training of the character, and English in a broad sense. Moreover, the barriers between individual subjects should be undermined in favor of an integrated approach to the common purposes of the curriculum and the pupils' needs as a whole. It also took the opportunity to address criticisms that the secondary school curriculum was out of touch with "real life": "The battle of Naseby [in the 1640s] may be more real than the machinery of Local Government. It is for the teacher to re-think the content and method of teaching existing subjects if reality is to be obtained." At the same time, Norwood was careful to ask each member of the committee to give their views on the advisability of replacing the external examination by an internal examination, clearly a device to flush out potential opposition, and it was noted that "all expressed opinions in favour of such a change provided that the safeguards which they considered necessary could be satisfactorily devised and operated."[62]

This set of decisions effectively gave Norwood a mandate to tell Butler that the most difficult stage had been passed, "not only in detail, but towards getting a common outlook on the future," and he was now confident of securing an agreed report, "a result of which hitherto I have been very doubtful."[63] Moreover, he was now clear on the lines on which the report would give its recommendations. He noted privately in a long letter to George Turner, his successor at Marlborough College, that the report would give a period of at least five years to develop the School Certificate as a record of work done and standards reached, to be followed by a permanent system under which schools would construct their own examinations with suitable external assessment. Examinations would not be held until the age of 18, at which point there would also be six months of national service for all. According to Norwood, this would be as significant for the future of teachers and teaching as it was for the pupils: "I want to give to the schoolteachers freedom and responsibility and the same right as that of the Universities to judge their own product: not otherwise can schoolmastering rise to the dignity of a profession." All of this, he acknowledged, would shock the IAHM, but he would now take on the professional bodies and try to "rope them in."[64] That is, as he more diplomatically expressed it to Butler, the committee still needed "to consult the Professional Bodies, and to work out carefully the mechanism of the changes which we propose over the coming years."[65]

Despite his growing confidence, his committee still managed to give him two setbacks before the report was finally agreed. First, it took objection to the term "national service" and preferred instead the more neutral phrase "an educational break."[66] Second, Terry Thomas, the headmaster from Leeds Grammar School, expressed his dissent at the final stage from the principle of the Internal Examination. Norwood made clear his exasperation at this late defection: "As our minds have been running in this direction for the last twelve months, he is somewhat belated in the discovery of his own mind, but I do not doubt that pressure has been put on him by the Association of Headmasters." Nevertheless, he took comfort that four other active teachers on the committee had sided with himself.[67] Butler was certainly satisfied with the fruits of Norwood's labor, deciding to have it published soon after the Education Bill and prior to a parliamentary debate to help inform the public about the issues.[68] According to Butler, "This well-written report will serve our book very well—particularly its layout of the Secondary world. Spens will be furious."[69]

The final Report was in many respects a distillation of Norwood's ideas on the nature and development of secondary education that he had been promoting for the preceding forty years. It began on the title page with a quotation from Plato's *Laws*, in the original Greek although with the concession of an English translation underneath. The inspiration of Plato was present throughout the report, and most evidently in Part One, which was what Butler referred to as its "layout of the Secondary world." This attempted to explain the key principles that underlay secondary education, and the different types of secondary school that should be developed in the future. It related these to a historical interpretation of education in England in which "rough groupings" of pupils had become recognized and treated in distinctive ways. Overall, it identified three different types of pupil, those associated with the grammar schools, those linked with technical schools, and those who would be most properly concerned with the modern schools. To each of these three ideal types it attached a particular approach to the curriculum. Thus, the grammar school pupil was

> the pupil who is interested in learning for its own sake, who can grasp an argument or follow a piece of connected reasoning, who is interested in causes, whether on the level of human volition or in the material world, who cares to know how things came to be as well as how they are, who is sensitive to language as expression of thought, to a proof as a precise demonstration, to a series of experiments justifying a principle; he is interested in the relatedness of related things, in development, in structure, in a coherent body of knowledge.[70]

Such pupils, it averred, should be catered to through a curriculum that treated fields of knowledge as suitable for "coherent and systematic study for their own sake apart from immediate considerations of occupation."[71] This could best be provided in a grammar school. Similarly the technical type of mind, concerned above all with the application of knowledge, would respond best to a technical curriculum in a technical school. The remainder, who formed the majority, were happier with "concrete things" rather than with ideas and should be provided with a broad curriculum that made a direct appeal to their interests in a modern school.[72] This was the classic Platonic hierarchy of knowledge in terms of gold, silver, and copper, applied to secondary education in the twentieth century. It did not elaborate on the social class implications as had the Taunton Report and Matthew Arnold in the 1860s. Nevertheless, these could be mapped out only too easily in a similar way, with the grammar schools as the locus for the middle classes and the modern schools as the natural repository for the children of the workers.

One key aspect that was missing from this graphic explication of educational differences was the position of the public schools. These had been located as potentially part of a federal system when the Norwood committee had begun its work in 1941 but were conspicuous by their absence from the final report. A separate committee had been established in 1942 to deal with these, it produced a report in 1944, but the opportunity that Norwood had sought to develop the connections between the State and the public schools was effectively lost.[73] To a large extent, this also signified the failure of Norwood's own personal project over the previous thirty years or more to reconcile the two sectors of education. The rift that had emerged between them, and which Norwood had struggled to heal, was to be formalized and deepened still further in the years to come.

The ways in which the three different kinds of secondary schools were to be distinguished from each other were, nonetheless, consistent with the approach that Norwood had favored since the publication of the Hadow Report in the 1920s. In classifying pupils for the different schools, it argued strongly in favor of using the "judgement of the teacher" in the primary schools, based on a record of the pupil's individual development "compiled by teachers who have known and taught him."[74] In a concession to the psychology of individual differences, it added that this could be supplemented, if desired, by intelligence tests, but the emphasis remained on the role of the teacher. It would be possible to transfer to a different kind of school at the age of 13, but again this would be based on a process in which "time and opportunity would be given for study of the relevant considerations, rather than a snap judgment dependent upon performance in an examination."[75] Unlike the Spens Report, it did not place great store by the notion of promoting parity of esteem between the different kinds of schools, recognizing that parity of conditions was not the same thing as winning recognition for a school.[76] In spite of this unapologetic justification of differentiating between pupils, it insisted that there were basic purposes common to the different types of curriculum and school, for the development of citizens as members of a community.[77] Like the typology of tripartism itself, this ideal of common citizenship was an uncritical adaptation of Platonic thought to the needs of modern society.

The following sections of the report spelled out implications for examinations and then for the curriculum along the lines already discussed. In Part Two, it recommended that the School Certificate examination should be reformed to become internal rather than external in nature, conducted by teachers at the school on syllabuses and papers framed by themselves. As a transitional measure over seven years, the examination would continue to be carried out by existing university-based examining bodies, but it would be reformed over that time, with a further decision made then as to whether to change to a wholly internal examination at that point. Part Three, the longest section of the report, pontificated on the curriculum and its general principles and aims before going on to examine the separate subjects in detail.

It was indeed a triumphant moment for Norwood and his support when he met his committee for the final time to sign off the report. Norwood was already active in explaining the report to the editors of key periodicals including the *Times Educational Supplement* and the *Journal of Education*, and he was satisfied that it had "made a good impression to start with." At the final committee meeting, he reverted to type "to assume the part of Head Master, and hold a Prize Giving, like the last day of term at a Public School."[78] He himself was presented on this occasion with a copy of the Cambridge Ancient History, "duly acknowledging the contribution of an Oxford man to future history through his leadership and tact."[79] This was a highly complacent celebration of a work that made few concessions either to the critics of the ingrained inequalities of the education system, or to the vested interests of the examination boards. The report was published in July 1943, shortly after the Education White Paper *Educational Reconstruction* and was thus central in paving the way for the reforms that were to follow.

Old War-Horses

The confidence of the Norwood Committee soon began to appear highly premature as vociferous objections were raised against the principles and recommendations in all three sections of the report. Indeed, over the following two years the report was widely criticized. The tripartite dimensions of secondary education did influence many LEAs in developing their plans for the future, but these also became highly controversial. The most obvious legacy of the report's main author, Norwood himself, was a divided system.

The proposals on the secondary school curriculum in Part Three of the Norwood Report attracted criticism from a number of interest groups. One such was the Council for Curriculum Reform, whose chairman, J.A. Lauwerys, announced his displeasure that "the sections which deal with the Curriculum have fallen flat." According to Lauwerys, "nearly every one expresses disappointment or a broad lack of interest" in relation to this aspect of the report. In part, this was because the report was concerned only with the curriculum of the established secondary schools rather than with those of the new types of schools that were to be developed. More broadly, he complained, the ideas on the curriculum that were supported in the report seemed vague and old-fashioned. They also appeared to lack specialist input from psychologists and sociologists.[80] Norwood responded to these complaints in his usual spirited manner, but he received only limited support.[81] Nor did the views of his report on individual subjects escape stricture. For example, the English Association expressed "disappointment and concern" with the section dealing with English, which it regarded as based on poorly founded criticisms and leading to reactionary recommendations. In particular, it was hostile to what it described as the report's "studied and unwarranted depreciation of the role of the specialist teacher of English."[82] Once again, Norwood was moved to defend himself strenuously, pointing out that the report had in fact emphasized the importance of the English language, and lamenting "faulty practice, and perhaps that lack of vision and ordinary fairness which make possible such a letter as that from the English Association."[83] There was in this reply a note of frustration that reflected a rapid fall from hubris to nemesis.

These snipings over the curriculum paled by comparison with the bitter conflict that ensued in relation to the Norwood Report's recommendations for examinations. The university examination boards in particular had strong reservations about the idea of moving toward internal examinations, and some of them carried these objections to the point of open resistance. Norwood claimed, indeed, that the Oxford Locals Delegacy and the Cambridge Local Syndicate were "doing their best to undermine and be-little the Report from pure motives of self-interest."[84] He argued strongly that the report should not need to report back to the SSEC since it had reported directly to the president of the Board of Education, but it was obliged to account for itself at a special meeting of the SSEC in November 1943. An unfriendly witness, James Petch of the Joint Matriculation Board, later recalled that Norwood's unwillingness to report back to the SSEC was tantamount to "official chicanery," and that this special meeting made little progress: "On assembly the members were bluntly informed hat their part was to receive and not to question; when it began to appear that considerable comment was likely, the Chairman [Norwood] unceremoniously dismissed them after only two hours."[85] Norwood had a very different view, as he told Butler after the meeting,

> We have to recognise that there is a definite attempt in progress on behalf of the Oxford Locals, the Cambridge Locals, and the Oxford and Cambridge Joint Board, to sabotage the report, and to throw the older universities into strong opposition. . . . The Secretary of the Cambridge side of the Oxford and Cambridge Joint Board had the effrontery to vote that our Report be *not* received—this after refusing to give evidence, or to answer our questions. But this Board has long ceased to play any useful part in our system.[86]

It was recorded in the minutes that the SSEC had agreed "with one dissentient" to receive the report,[87] but clearly this did not begin to reflect the hostility to which it had given rise.

Norwood continued to hope that the School Certificate examination in its current form would eventually be abandoned just as he had suggested in May 1944: "My view is that the School Certificate will die of inanition, if it is left to itself."[88] Again, however, he paid too little heed to the conservatism of the secondary schools, of the teachers and head teachers, and of the universities, which resolutely held back from endorsing the radical reforms that he had championed. After protracted discussions, external examinations were maintained in the new guise of the General Certificate of Education (GCE)that was introduced as from 1950.[89]

In many ways, however, the most significant rebuffs to the Norwood Report were in relation to its general layout of secondary education and the basic typology of three types of secondary school. This went to the heart of the report's general role in the reconstruction of the education system. Norwood himself had viewed this in terms of restraining the radical impulses that the war and the debates over educational reform had served to encourage. The idea of a "New Order," he argued, was dangerous if it led to the loss of established values and institutions. It was necessary therefore to widen educational opportunity without, as he saw it, "destroying educational values." Indeed, he urged, "it would be a thousand pities if through the mistaken equalitarian enthusiasm of the moment we were to use education to undermine the primary values for which we are fighting this war."[90] In private he was still franker in voicing his fears that "in destroying privilege they destroy not only variety but excellence, not only license but liberty," and a conviction that the education of the future would fall victim to "the inevitable totalitarianism of the Left or of the Right."[91]

There was in these forebodings a strong element of the fear and anxiety that had guided Norwood's educational attitudes for at least the previous forty years. This helps to explain the rigidity of the types to which his report attached general labels and descriptions. If there was to be an expansion of educational opportunities, it would need perforce to be accompanied by the preservation of educational institutions that he had supported throughout his educational career. It was a response on his part that emphasized the continuity of cultural values and traditions, but it ran completely against the prevailing political support for greater social equality. This political movement was consolidated at the end of the war in July 1945, with the election of a Labour government with a large parliamentary majority under Clement Attlee.

Norwood's conservatism in this respect represented a social class project on behalf of the middle classes. He had tried but failed to find a way to unite the middle classes by bringing the public schools and grammar schools together as Matthew Arnold had proposed. Now, *faute de mieux*, he was attempting to maintain the distinctions that surrounded the grammar schools, as a superior and traditional form of provision in the new world of secondary education for all. It was this that was most distasteful to Fred Clarke, the director of the Institute of Education in London. Although careful to be diplomatic in public, he was privately scathing in his attack on the report's general approach, which he viewed as "Norwood at his worst." According to Clarke, "The thing is just clogged up with typical public-school woolliness, and all English provincialism of the blindest sort." He feared that it would damage the reputation of the new English system, once it circulated in other countries. Finally, he declared, "People like Norwood imagine themselves to be talking principles when what is really moving them is just uncritical insular prejudice, and class-prejudice at that. His sort may be a grave danger to us, and I begin to feel that they are quite unteachable."[92] This kind of reaction against Norwood's ideals was soon to become prevalent, and Norwood himself was cast as the representative of the *ancien regime*. It is true that in many cases the LEAs that had adopted a tripartite approach to secondary education in their own areas were following their own interests and ideals rather than it being imposed upon them.[93] Nevertheless, it came to be widely regarded as Norwood's most lasting legacy to education.

Norwood's own direct role in the forming of education policy was also nearing its end. He recommended that the SSEC, over which he had presided for nearly a quarter of a century, should now be wound up to "share an honourable grave with the Consultative Committee." It was, he suggested, "a body which can supervise the running of an established system, but is incapable of planning a new one."[94] He was himself an obvious candidate to be invited to chair the new Advisory Council on Education that was established under the Education Act of 1944 to succeed the consultative committee. However, it was recognized that he was too old and too far out of sympathy with the postwar generation for this to appear to be a forward-looking development. One official who drew up a list of possible choices suggested the leading economist Lord Keynes as the chairman of the Advisory Council and acknowledged that this might seem odd. On the other hand, he explained, "I wished however to avoid choosing one of the old war-horses like Sir Will Spens or Sir Cyril Norwood and I could not think of anybody who was sufficiently distinguished and yet can be counted upon to take an intelligent interest in the subject."[95] Butler agreed that these figures were now too elderly to chair the new body, although he still held out for "some representatives of the old school" to be included in its membership, "in view of the need for continuity."[96] In the event, it was to be Fred Clarke, of a similar vintage but more in sympathy with the social aspirations of the educational reforms, who became the first chairman of the Advisory Council of Education.

This left only the SSEC itself. When Ellen Wilkinson became minister of education for the Labour government in 1945, Norwood's days as its chairman were clearly numbered. The Board of Education officials who had worked closely with him during the war made a case for retaining his services. Williams in particular emphasized his experience and proposed that he be invited to continue as chairman, adding that "I believe he would be prepared to do this for another two or three years."[97] Holmes, retiring as permanent secretary, expressed his support for this suggestion.[98] But it was Holmes himself who was invited to take over as the chairman, and the minister duly thanked Norwood for his services over the past twenty-five years.[99] It was ironic, yet somehow fitting, that Norwood's educational career at a national level should come to a close in this way. In 1917, he had been opposed by the board officials but was picked out by a reforming minister, Herbert Fisher. Now the officials were on his side, but the new political broom had him targeted as an outmoded survivor from the past. It was one more indication of the contested nature of Norwood's educational career and of his general contribution to educational policy and ideas—a man who in his parting, just as in his arrival, polarized opinion and never quite established the general approval of his peers.

Chapter 10

Conclusions: The Ideal of Secondary Education

This book has studied in depth the development of secondary education over eighty years of change, from Matthew Arnold and the Taunton Report in the 1860s to World War II. Over this time, secondary education was intended to provide an advanced form of education that would be suited to prepare pupils for universities and the professions. This basic purpose gave rise to fierce debates over how best to carry out such a task that went to the heart of class relationships, values, and tradition.

In investigating these debates, the current work has sought to build especially on the previous contributions of Banks and Simon.[1] Banks established a persuasive framework of analysis that emphasized the vocational aspects of secondary education in a differentiated employment market; Simon was no less influential in his graphic depiction of social class inequalities. In the discussion developed here, the broad range of the middle classes has been the starting point of an exploration of the tensions between grammar schools and public schools. It has also attempted to address what Simon described as the "staying power" of secondary education, which was rooted in the emotions just as much as the calculations of social class. Secondary education drew on academic and social aspirations but was at the same time a defensive set of projects. The public schools tended to retreat from contamination from the grammar schools. Similarly, the grammar schools tried to maintain a superior status in relation to newer forms of education that might threaten their social position. Within this broader context, the curriculum, examinations, and teachers themselves were in general conservative and highly resistant to reform. In order to understand these tensions and issues more deeply, the book has combined a case-study approach to particular institutions with a detailed biographical treatment of one of the major figures of the time. This approach has helped to gain purchase on the experience of education during these years, while also allowing an engagement with key texts, national policy developments, and the broader international context.

Deeply immersed in these debates was Cyril Norwood, who dominated a generation of educational reform and change in secondary education in the early twentieth century. More than his contemporaries such as R.H. Tawney or Cyril Burt, he defined the ideal of secondary education and lived it as a teacher and a headmaster and also as a policymaker and a polemicist for more than forty years. He was in many ways very limited in his outlook and contribution. He certainly did not possess the philosophical rigor of John Dewey, the sociological insights of Emile Durkheim, the comparative understandings of Michael Sadler, the historical depth of Tawney, the psychological expertise of Burt, nor yet the administrative talent of C.E. Beeby.[2] He was an outstanding scholar with a broad and enquiring mind, though his keenest interests were in history, the classics, and English rather than in science and technology. His involvement in secondary education covered a wide compass, rather than being profound in its depth by focusing on a particular topic or problem.

Norwood had a fundamentally conservative frame of mind that favored social hierarchy, established institutions, and ingrained traditions. He was nationalistic and an imperialist in his attachment to the trappings of the empire. At the same time, he was radical in his views about how to adapt these institutions to the changing needs of twentieth-century society. He was keen to promote the interests of teachers as a professional group, to challenge the power of examinations, and to promote the idea of citizenship. He could be iconoclastic in his public utterances and was at times isolated and exposed in his commitment to particular aims. Most fundamentally of all, perhaps, he saw in the State the potential to generate reform and to resolve deep tensions. This was a complex pattern of beliefs and allegiances, one that allowed him to speak to a range of diverging interests. In the end, however, he failed to gain lasting support for shared objectives. His conservatism increasingly alienated critics of the system, while his reforming agenda was too alarming for conservatives and established interest groups.

In terms of a positive or creative goal, the key project of secondary education with which Norwood was associated was to fulfill Arnold's vision of uniting the middle classes in day schools and boarding schools through the good offices of the State. During this period he crossed the divide between these different kinds of schools, preaching the importance of unity. Despite his efforts this project failed, and indeed the gap between these different types of school became wider and more difficult to traverse. After World War II, the public schools became the preferred alternative for middle-class parents who feared the provision of secondary education under the auspices of the State. Indeed, there developed an increasingly antagonistic relationship between the State and the private sector that led to them being widely regarded as "two nations." The divide between them symbolized inequalities in status and social class that continued to be evident into the twenty-first century.[3]

Norwood was also deeply involved in the defensive project to maintain what he and his influential allies regarded as the integrity of secondary education against the challenge of democracy, and to manage its survival against the encroachment of rival forms that might undermine established values and institutions. This also failed ultimately, as the grammar schools were eclipsed from the 1960s onward. He came to be vilified as the symbol of the *ancien regime*, even though grammar schools were to be venerated by some as a continuing challenge to the development of more egalitarian forms of secondary education.

Norwood himself represented a particular kind of middle-class figure, with clerical and provincial roots in the nineteenth century. This was a declining species in the context of the more scientific, metropolitan, meritocratic middle class that emerged in the 1930s and 1940s. By the time of World War II and thereafter, religion was becoming eclipsed by secular concerns that emphasized particular social and economic issues. In this changing climate, Norwood himself came to appear very much a back number.

The insecurity and fear of the middle classes were pervasive throughout this period. Norwood himself was a very good example of this through his own class and family background. He could be regarded as an outsider, socially just as much as politically, especially in the elite institutions of Marlborough and most particularly Harrow. Fear of the future was characteristic not only of the *English Tradition of Education* in 1929 but also of the Norwood Report—the ultimate and defiant expression of a defensive project against democracy and equality. It reflected vividly the rigidity and fundamentalism to which its principal author had become increasingly prone, as well as Norwood's taste for confrontational public utterances.

And yet, in its time, the ideal of secondary education that Norwood represented was potent in its attraction. In his stewardship of Bristol Grammar School and in his national role in education policy, he provided a rhetoric that sustained the position of grammar schools and a particular kind of curriculum. He helped to give it the "staying power" that became so characteristic, an allure that defeated its rivals. Moreover, the ideal of secondary education that was current before World War II has resonance even in the debates of our own time. The debates over secondary education that have flared up again at the end of the twentieth and the start of the twenty-first century demonstrate that the purposes and ideals of secondary education have not been resolved, that the struggles over curriculum, examinations, and teachers that were so fierce in Norwood's day have far from run their course. Memories of grammar schools and public schools continue to play a part in these debates and help to explain the passion with which they have been fought. The emotions of social class that underlay secondary education during the late nineteenth and early twentieth centuries have again become clearly exposed to view. The controversies surrounding the Education White Paper of 2005 have dwelt on the way in which the middle classes engage with the State in the sphere of secondary education, and the political dominance of the "pushy middle classes."[4] Half a century and more after Norwood's death, and over a century after Durkheim pointed out the long-term dilemmas involved, the ideal of secondary education remains fiercely contested.

To begin to understand why this is so, as Durkheim so eloquently explained, we must look first to our history. It is as superficial as it is misleading to address the continuing problems of secondary education as though they are only of our own time and of the present generation. A fuller and longer-term conceptualization of the tensions and the ideals involved is a prerequisite to comprehending both the continuing conflicts over secondary education, and the challenges of the future.

Notes

Chapter 1   Introduction: Cyril Norwood and Secondary Education

Notes to Pages 1–6

1. Emile Durkheim, *The Evolution of Educational Thought: Lectures on the Formation and Development of Secondary Education in France* (London, RKP, 1977), p. 8.
2. Ibid., p. 10.
3. Ibid., p. 13.
4. C. Wright Mills, *The Sociological Imagination* (London, Oxford University Press, 1959), p. 5.
5. Ibid., p. 6.
6. Ibid., p. 7.
7. Ibid.
8. R.J.W. Selleck, *James Kay-Shuttleworth: Journey of an Outsider* (London, Woburn, 1994), p. xiv.
9. J. Goodman and J. Martin, *Women and Education, 1800–1980* (London, Palgrave Macmillan, 2004), p. 6.
10. G.C. Turner, "Norwood, Sir Cyril (1875–1956)," *Dictionary of National Biography*, 1951–1960, p. 773.
11. See also Gary McCulloch, "From Incorporation to Privatisation: Public and Private Secondary Education in Twentieth-Century England," in Richard Aldrich (ed.), *Public or Private Education?: Lessons from History* (London, Woburn, 2004), pp. 53–72; Gary McCulloch, "Cyril Norwood and the English Tradition of Education," *Oxford Review of Education*, 32/1 (2006), pp. 55–69; and Gary McCulloch, "Education and the Middle Classes: The Case of the English Grammar Schools, 1868–1944," *History of Education*, 35/6 (2006), pp. 689–704.
12. John Graves, *Policy and Progress in Secondary Education, 1902–1942* (London, Thomas Nelson, 1943), p. viii.
13. Olive Banks, *Parity and Prestige in English Secondary Education: A Study in Educational Sociology* (London, Routledge and Kegan Paul, 1955), p. 12.
14. Ibid., p. 239.
15. Ibid., p. 241.
16. Brian Simon, *The Politics of Educational Reform, 1920–1940* (London, Lawrence and Wishart, 1974), p. 10.
17. Ibid.
18. Ibid., p. 318.
19. Ibid., p. 333.

20. Felicity Hunt, *Gender and Policy in English Education: Schooling for Girls, 1902–44* (London, Harvester Wheatsheaf, 1991).
21. Ibid., p. 1.
22. R.H. Tawney, *Equality* (London, George Allen and Unwin, 1931; 1964 edition), pp. 144–45.
23. Brian Simon, "The 1902 Education Act—A Wrong Turning," *History of Education Society Bulletin*, 70 (2002), p. 74.
24. Mel Vlaeminke, "Supreme Achievement or Disastrous Package? The 1902 Act revisited," *History of Education Society Bulletin*, 70 (2002), p. 76.
25. Ibid., p. 85.
26. Mel Vlaeminke, *The English Higher Grade Schools: A Lost Opportunity* (London, Woburn, 2000), p. 29.
27. Kevin Manton, *Socialism and Education in Britain, 1883–1902* (London, Woburn, 2001), p. 197.
28. Vlaeminke, *The English Higher Grade Schools*, chapters 3 and 4.
29. Manton, *Socialism and Education in Britain*, esp. chapter 3.
30. See, e.g., Gary McCulloch, *Educational Reconstruction: The 1944 Education Act and the Twenty-First Century* (London, Woburn, 1994).
31. Board of Education, *Curriculum and Examinations in Secondary Schools* (Norwood Report) (London, HMSO, 1943), p. vii.
32. Ibid.
33. Ibid., p. viii.
34. Ibid., p. ix.
35. For example, Brian Simon, "The 1944 Education Act: A Conservative measure?" *History of Education*, 15/1 (1986), pp. 31–43; McCulloch, *Educational Reconstruction*, esp. chapters 4, 5.
36. For a recent assessment of the historical development of schools in the United States see William J. Reese, *America's Public Schools: From the Common School to "No Child Left Behind"* (Baltimore, Johns Hopkins University Press, 2005).

Chapter 2   Middle Class Education and the State

1. Fred Clarke, *Education and Social Change: An English Interpretation* (London, Sheldon Books, 1940), p. 10.
2. Ibid., p. 35.
3. For examples of a massive literature in this area, see Rupert Wilkinson, *The Prefects: British Leadership and the Public School Tradition* (London, Oxford University Press, 1964); J.R. de S. Honey, *Tom Brown's Universe: The Development of the Public School Community in the 19th Century* (London, Millington Books, 1977); J.A. Mangan, *Athleticism in the Victorian and Edwardian Public School* (London, Falmer, 1986); Brian Simon, *The Two Nations and the Educational Structure, 1780–1870* (London, Lawrence and Wishart, 1960); Brian Simon, *The Politics of Educational Reform, 1920–1940* (London, Lawrence and Wishart, 1974). Two of my earlier works, *Philosophers and Kings: Education for Leadership in Modern England* (Cambridge, Cambridge University Press, 1991) and *Failing the Ordinary Child? The Theory and Practice of Working Class Secondary Education* (Buckingham, Open University Press, 1998) explored these "elite" and "mass" traditions respectively.

4. Peter Searby, "Foreword," in P. Searby (ed.), *Educating the Victorian Middle Class* (Leicester, History of Education Society, 1982), p. vi.
5. See esp. W.E. Marsden, *Unequal Educational Provision in England and Wales: The Nineteenth-Century Roots* (London, Woburn, 1987); W.E. Marsden, *Educating the Respectable: A Study of Fleet Road School, Hampstead, 1879–1903* (London, Woburn, 1991); David Reeder, "The Reconstruction of Secondary Education in England, 1869–1920," in D. Muller, F. Ringer, and B. Simon (eds.), *The Rise of the Modern Educational System: Structural Change and Social Reproduction 1870–1920* (Cambridge, Cambridge University Press, 1987), pp. 133–50.
6. Tawney, *Equality*, p. 64.
7. Ibid.
8. Joanna Bourke, *Fear: A Cultural History* (London, Virago, 2005), p. 27.
9. Burton J. Bledstein, "Introduction: Storytellers to the Middle Class," in B.J. Bledstein and R.D. Johnston (eds.), *The Middling Sorts: Explorations in the History of the America Middle Class* (London, Routledge, 2001), p. 5.
10. Stanley Aronowitz, *How Class Works: Power and Social Movement* (New Haven, Yale University Press, 2003), p. 34.
11. Peter Gay, *The Bourgeois Experience, Victoria to Freud*, vol. 1, Education of the Senses (Oxford, Oxford University Press, 1984), pp. 17, 67.
12. Mills, *The Sociological Imagination*, p. 5.
13. Vlaeminke, *The English Higher Grade Schools*; Manton, *Socialism and Education*.
14. A.H. Halsey, "The Relation between Education and Social Mobility with Reference to the Grammar School since 1944" (PhD thesis, University of London, 1954).
15. Felicity Hunt, "Social Class and the Grading of Schools: Realities in Girls' Secondary Education, 1880–1940," in June Purvis (ed.), *The Education of Girls and Women* (Leicester, History of Education Society, 1985), pp. 27–46.
16. Barry Blades, "Deacon's School, Peterborough, 1902–1926: A study of the Social and Economic Function of Secondary Schooling" (PhD thesis, University of London, 2003).
17. F.M.L. Thompson, *The Rise of Respectable Society: A Social History of Victorian Britain, 1830–1900* (London, Fontana, 1988)
18. Ross McKibbin, *Classes and Cultures: England 1918–1951* (Oxford, Oxford University Press, 1998), p. 104.
19. See, e.g., Alan Kidd and David Nicholls (eds.), *The Making of the British Middle Class? Studies of Regional and Cultural Diversity since the Eighteenth Century* (Stroud, Sutton Publishing, 1998); and Alan Kidd and David Nicholls (eds.), *Gender, Civic Culture and Consumerism: Middle Class Identity in Britain 1800–1940* (Manchester, Manchester University Press, 1999).
20. Leonore Davidoff and Catherine Hall, *Family Fortunes: Men and Women of the English Middle Class, 1780–1950* (London, Hutchinson, 1987), pp. 28, 35. See also, e.g., R.J. Morris, "A Year in the Public Life of the British Bourgeoisie," in Colls and Rodger (eds.), *Cities of Ideas*, pp. 121–43
21. Geoffrey Crossick, "From Gentlemen to the Residuum: Languages of Social Description in Victorian Britain," in Penelope Corfield (ed.), *Language, History and Class* (London, Basil Blackwell, 1991), p. 173.
22. For example, Richard Trainor, "The Middle Class," in Martin Daunton (ed.), *The Cambridge Urban History of Britain*, vol. III, 1840–1950 (Cambridge, Cambridge University Press, 1991), pp. 673–713.
23. Fiona Devine and Mike Savage, "The Cultural Turn: Sociology and Class Analysis," in Fiona Devine, Mike Savage, John Scott, and Rosemary Crompton (eds.), *Rethinking Class: Culture, Identities and Lifestyles* (London, Palgrave Macmillan, 2005), pp. 1–23; Beverley

Skeggs, *Class, Self, Culture* (London, Routledge, 2004). See also Diane Reay, "Thinking Class, Making Class," *British Journal of Sociology of Education*, 26/1 (2005), pp. 139–45.
24. For example, Fiona Devine, *Class Practices: How Parents Help Their Children to Good Good Jobs* (Cambridge, Cambridge University Press, 2004). See also Andy Furlong, "Making Middle Class Advantage," *British Journal of Sociology of Education*, 26/5 (2005), pp. 683–85.
25. See, e.g., Michael Grenfell, *Bourdieu and Education: Acts of Practical Theory* (London, Falmer, 1998); and Nicholas Brown and Jane Szeman (eds.), *Pierre Bourdieu: Fieldwork in Culture* (Lenham, MA; Renman and Littlefield, 2000).
26. Sally Power et al, *Education and the Middle Class* (Buckingham, Open University Press, 2003).
27. Stephen Ball, *Class Strategies and the Education Market: The Middle Classes and Social Advantage* (London, Routledge Falmer, 2003), p. 2.
28. Ibid., p. 4.
29. Sally Tomlinson, *Education in a Post-Welfare Society* (2nd edition, Maidenhead, Open University Press, 2005), p. 173.
30. Report of Schools Inquiry Commission (Taunton Report) (1868), vol. I, pp. 15–16.
31. Ibid., pp. 17–18.
32. Ibid., p. 20.
33. Ibid., p. 44.
34. Report of a Conference on Secondary Education in England, October 10–11, 1893 (Oxford, Clarendon Press, 1893), p. 15.
35. Royal Commission on Secondary Education (1895) (Bryce Report), Part III, p. 136.
36. Ibid., p. 138.
37. See, e.g., Felicity Hunt, "Divided Aims: The Educational Implications of Opposing Ideologies in Girls' Secondary Schooling, 1850–1940," in Felicity Hunt (ed.), *Lessons for Life: The Schooling of Girls and Women, 1850–1950* (London, Basil Blackwell, 1987), pp. 3–21; and A.M. Wolpe, "The Official Ideology of Education for Girls," in M. Flude and J. Ahier (eds.), *Educability, Schools and Ideology* (London, Croom Helm), pp. 138–59.
38. Joyce Senders Pedersen, *The Reform of Girls' Secondary and Higher Education in Victorian England: A Study of Elites and Educational Change* (London, Garland, 1987), chapter 9.
39. John Roach, *Secondary Education in England 1870–1902: Public Activity and Private Enterprise* (London, Routledge, 1991), p. 229.
40. Taunton Report, p. 18.
41. Matthew Arnold, "Schools and Universities on the Continent," in R.H. Super (ed.), *Schools and Universities on the Continent* (Ann Arbor, University of Michigan Press, 1964); first written as a report to the Taunton Commission, p. 35.
42. Ibid., p. 304.
43. Ibid., p. 309.
44. Ibid.
45. Ibid., p. 314.
46. Ibid., p. 328.
47. Matthew Arnold, "A French Eton, or Middle-Class Education and the State," in Peter Smith and Geoffrey Summerfield (eds.), *Matthew Arnold and the Education of a New Order* (Cambridge, Cambridge University Press, 1969), p. 98.
48. Ibid., p. 112.
49. Ibid., pp. 112–13.
50. Ibid., pp. 115–16.
51. Ibid., p. 117.
52. Ibid., p. 120.
53. Ibid., p. 132.

54. Ibid., p. 145.
55. Ibid., p. 152.
56. Matthew Arnold, "Culture and Anarchy," in J. Dover Wilson (ed.), *Culture and Anarchy* (1869) (Cambridge, Cambridge University Press, 1932), p. 27.
57. Ibid., p. 70.
58. Ibid.
59. Ibid., p. 71.
60. Ibid., p. 94.
61. Ibid., p. 95.
62. Ibid., p. 96.
63. See, e.g., Frank M. Turner, "Why the Greeks and not the Romans in Victorian Britain?," in G.W. Clarke (ed.), *Rediscovering Hellenism: The Hellenic Inheritance and the English Imagination* (Cambridge, Cambridge University Press, 1989), pp. 61–81; and James Bowen, "Education, Ideology and the Ruling Class: Hellenism and English Public Schools in the Nineteenth Century," Ibid., pp. 161–86.
64. John Roach, *A History of Secondary Education in England, 1800–1870* (London, Longman, 1986), p. 261.
65. See also, e.g., T.W. Bamford, *Thomas Arnold* (London, The Cresset Press, 1960), and Michael McCrum, *Thomas Arnold, Head Master: A Reassessment* (Oxford, Oxford University Press, 1989), for further details.
66. McCrum, p. 25.
67. Ibid., pp. 5–8; Bamford, chapter 7.
68. Thomas Hughes, *Tom Brown's Schooldays* (1857 / London, Dent, 1949), p. 51.
69. Ibid., p. 65.
70. Ibid., p. 78.
71. Ibid., p. 127.
72. J.R. de S. Honey, *Tom Brown's Universe: The Development of the Public School in the 19th Century* (London, Millington Books, 1977); see also, e.g., Rupert Wilkinson, *The Prefects: British Leadership and the Public School Tradition* (London, Oxford University Press, 1964).
73. W.E. Marsden, "Schools for the Urban Lower Middle Class: Third Grade or Higher Grade?," in Peter Searby (ed.), *Educating the Victorian Middle Class* (Leicester, History of Education Society, 1982), p. 55.
74. Vlaeminke, *The English Higher Grade Schools*.
75. See David Reeder, "The Reconstruction of Secondary Education in England, 1869–1920," in D. Muller, F. Ringer, and B. Simon (eds.), *The Rise of the Modern Educational System: Structural Change and Social Reproduction, 1870–1920* (Cambridge, Cambridge University Press, 1987), pp. 135–50; and Brian Simon, "David Reeder's 'Alternative System': The School Boards in the 1890s," in R. Colls and R. Rodger (eds.), *Cities of Ideas: Civil Society and Urban Governance in Britain, 1800–2000* (Aldershot, Ashgate, 2004), pp. 178–206.
76. Charles Dickens, *The Personal History of David Copperfield* (1850/1948, Oxford, Oxford University Press), p. 77.
77. Ibid.
78. Ibid., p. 78.
79. Ibid., p. 86.
80. Ibid.
81. Ibid., p. 87.
82. Ibid., p. 99.
83. Ibid.

84. Charles Dickens, p. 100.
85. Ibid., p. 101.
86. Ibid., p. 94.
87. Ibid., p. 95.
88. Ibid., p. 226.
89. Ibid., p. 237.
90. Ibid.
91. Ibid., pp. 237–38.
92. See also, e.g., Gary McCulloch, *Documentary Research in Education, History and the Social Sciences* (London, RoutledgeFalmer, 2004).
93. F.G. Walcott, *The Origins of Culture and Anarchy: Matthew Arnold and Popular Education in England* (Toronto, Toronto University Press, 1970), p. 113.

Chapter 3   The Education of Cyril Norwood

1. John Gough Nichols, Ponsonby A. Lyons, *A History of the General Parish of Whalley* (4th edition, Routledge, 1876), vol. 11, pp. 17–18.
2. Ibid., p. 18.
3. Revd Samuel Norwood to the Endowed Schools Commission, August 11, 1870 (National Archives, ED.27/2318).
4. Ibid.
5. See also, e.g., David Allsobrook, *Schools for the Shires: The Reform of Middle-Class Education in Mid-Victorian England* (Manchester, Manchester University Press, 1986), esp. chapters 8–10.
6. Endowed Schools Commission to Revd Samuel Norwood, October 16, 1870 (National Archives, ED.27/2318).
7. Report by Mr J. Bryce (Schools Inquiry Commission, vol. IX, General Reports by Assistant Commissioners: Northern Counties), p. 428.
8. Ibid., p. 429.
9. Ibid., p. 491.
10. Ibid.
11. Ibid.
12. Ibid.
13. Ibid., p. 493.
14. Ibid., p. 494.
15. Ibid., pp. 690–91.
16. Ibid., p. 694.
17. Ibid., p. 697.
18. Ibid., p. 497.
19. Ibid., p. 498.
20. Ibid.
21. Ibid., pp. 430, 491.
22. Ibid., vol. 17, Whalley Grammar School: Mr Bryce's Report, p. 430.
23. Ibid., p. 431.
24. Ibid.
25. Ibid., p. 432.

26. Ibid.
27. *Crockford's Clerical Directory* (1870), p. 521.
28. *Preston Guardian*, July 3, 1869, supplement, p. 3.
29. Rev. Samuel Norwood, *Our Indian Empire: The History of the Wonderful Rise of British Supremacy in Hindustan* (London, Samuel Tinsley, 1876), Preface.
30. Ibid., p. 345.
31. Ibid., pp. 345–46.
32. Sarah Norwood, Death certificate, Whalley, October 19, 1868.
33. Samuel Norwood and Elizabeth Emma Sparks, marriage certificate, Whalley, July 7, 1869.
34. See, e.g., Linda Young, *Middle-Class Culture in the Nineteenth Century: America, Australia and Britain* (London, Palgrave Macmillan, 2003).
35. R. Durnford, report to the Secretary, Endowed Schools Department, January 3, 1881 (National Archives, ED.27/2318).
36. Ibid.
37. Ibid.
38. Ibid.
39. Ibid.
40. Bryce Commission on secondary education, 1895, Appendix: return of foundations in England.
41. Clarendon Report (1864), vol. I, p. 202.
42. W.C. Farr (ed.), *Merchant Taylors' School: Its Origin, History and Present Surroundings* (London, Basil Blackwell, 1929), chapter 3.
43. Clarendon Report (1864), p. 225.
44. Ibid., vol. 2, p. 252.
45. Ibid., vol. 4, p. 117.
46. Ibid., vol. 2, p. 265.
47. These details are drawn from Farr (ed.), *Merchant Taylors' School* and F.W.M. Draper, *Four Centuries of Merchant Taylors' School, 1561–1961* (London, Oxford University Press, 1961).
48. Draper, p. 196.
49. Farr, p. 81.
50. Draper, pp. 161, 196.
51. Francis West, *Gilbert Murray: A Life* (London, Croom Helm, 1984), p. 14.
52. Ibid., p. 16.
53. *The Taylorian*, vol. XI (1889), p. 165.
54. Ibid., vol. XIII (1891), p. 183.
55. Ibid., vol. XVI (1894), p. 201.
56. Ibid., "The School at Charterhouse Square," Part 3, vol. LXIII (1948, no. 2), p. 109.
57. Ibid., vol. XV (1893), p. 28.
58. Ibid., vol. XVI (1894), p. 19.
59. Ibid., vol. XIV (1892), p. 133.
60. Ibid., p. 134.
61. Ibid., vol. XV (1893), pp. 94–95.
62. Ibid., p. 122.
63. Ibid., vol. XVI (1894), pp. 97–98.
64. Ibid., vol. XV (1893), p. 170.
65. William Baker, testimonial to the Governors of Bath College, March 27, 1902 (Norwood papers, University of Sheffield, 230/2/1).
66. See John G. Davies, *Leeds Grammar School: A Pictorial History* (Otley, 1991).
67. Leeds Grammar School, *Calendar* (1901).
68. Ibid.

69. Admiralty testimonial for Cyril Norwood to King Edward VII Grammar School, Sheffield, February 2, 1905 (Norwood papers, University of Sheffield, 230/2/3).
70. Correspondence with Admissions Registrar, Notting Hill and Ealing High School.
71. Cyril Norwood, letter to Catherine Kilner, February 5, 1897 (Norwood papers, 230/10/1B).
72. Cyril Norwood to Catherine Kilner, n.d. [January 1897] (Norwood papers, 230/10/1B).
73. Cyril Norwood, letter to Catherine Kilner, February 5, 1897 (Norwood papers, 230/10/1B).
74. Cyril Norwood to Catherine Kilner, November 17, 1901; Catherine Kilner to Cyril Norwood, December 14, 1901 (Norwood papers, 230/10/!A, 1B).
75. Catherine Kilner to Cyril Norwood, n.d. [late November 1901] (Norwood papers, 230/10/1A).
76. For example, Catherine Kilner to Cyril Norwood, February 4, 1897, April 14, 1897 (Norwood papers, 230/10/1A).
77. Catherine Kilner to Cyril Norwood, October 11, 1897, November 1, 1897 (Norwood papers, 230/10/1A).
78. Catherine Kilner to Cyril Norwood, March 3, 1898 (Norwood papers, 230/10/1A).
79. Catherine Kilner to Cyril Norwood, September 6, 1898 (Norwood papers, 230/10/1A).
80. Catherine Kilner to Cyril Norwood, August 27, 1897 (Norwood papers, 230/10/1A).
81. Catherine Kilner to Cyril Norwood, February 14, 1898 (Norwood papers, 230/10/1A).
82. *The Leodiensian*, vol. XXII, no. 4 (July 1903), p. 112.
83. Ibid., vol. XXV, no. 4 (July 1906), "Vale," p. 74
84. Ibid., vol. XXI, no. 3 (May 1902), p. 68.
85. Ibid., vol. XXII, no. 2 (April 1903), pp. 43–44.
86. Ibid., no. 3 (May 1903), p. 92.
87. Ibid., vol. XX, no. 6 (December 1901), p. 110.
88. Leeds Grammar School, headmaster's log book, October 20, 1905 (Leeds Grammar School papers).
89. *The Leodiensian*, vol. XXIV, no. 1 (February 1905), p. 12.
90. Ibid., vol. XXV, no. 4 (July 1906), p. 74.
91. Ibid., vol. XXIII, no. 2 (April 1904), "Manent ea fata nepotes," Part I, p. 22.
92. Ibid., p. 23.
93. *The Leodiensian*, vol. XXIII, no. 3 (June 1904), Part II, pp. 45–49.
94. Ibid., p. 49.
95. *The Leodiensian*, vol. XXII, no. 5 (October 1903), report on Speech Day, July 29, p. 120.
96. Inspection report, Leeds Grammar School, October 3–5, 1905 (National Archives, ED.109/7313).
97. Sir Cyril Norwood, letter to George Turner, September 27, 1942 (St. John's College Oxford papers).
98. A.T. Barton, letter to Governors of Bath College, April 1, 1902 (Norwood papers, 230/2/1); Admiralty reference letter for Cyril Norwood, February 2, 1905 (Norwood papers, 230/2/3).
99. Cyril Norwood, letter to the Governors of Bristol Grammar School, n.d. [1906] (Bristol Grammar School papers).
100. Samuel Norwood, diary, July 28, 1905 (Norwood papers, 230/11/2).
101. Samuel Norwood, diary, August 18, 1905 (Norwood papers, 230/11/2).
102. Samuel Norwood, diary, October 21, 1905 (Norwood papers, 230/11/2).
103. Samuel Norwood, diary, May 3, 1906 (Norwood papers, 230/11/2).
104. Samuel Norwood, diary, November 18, 1907 (Norwood papers, 230/11/2).

Chapter 4 The Higher Education of Boys in England

1. Details on the history of Bristol Grammar School are drawn from C.P. Hill, *The History of Bristol Grammar School* (London, Pitman, 1951). Further details on Bristol's higher grade schools are in Vlaeminke, *The English Higher Grade Schools*, esp. chapter 3.
2. R.L. Leighton, letter to BGS Governors, October 27, 1904 (BGS Governors Minute Book, Bristol Record Office).
3. Meeting of BGS Board of Governors, January 3, 1905 (BGS Governors Minute Book, Bristol Record Office).
4. *Bristol Grammar School Chronicle*, vol. IX, no. 7 (April 1905), editorial, p. 87.
5. Board of Education, report of inspection, Bristol Grammar School, March 30–April 1, 1905 (Board of Education papers, National Archives, ED.109/1599).
6. Ibid.
7. Meeting of BGS board of governors, November 3, 1905 (BGS Governors Minute Book, Bristol Record Office).
8. Ibid.
9. Meeting of BGS Board of Governors, November 24, 1905 (BGS Governors Minute Book, Bristol Record Office).
10. Special Meetings of BGS Board of Governors, March 17, 1906; April 4, 1906, April 24, 1906 (BGS Governors Minute Book, Bristol Record Office).
11. Thomas Methley, letter to Cyril Norwood, December 27, 1905 (Norwood papers, University of Sheffield, 230/2/4).
12. J.R. Gerans to Rev J.R. Wynne-Edwards, December 21, 1905 (Norwood papers, 230/2/4).
13. *Bristol Grammar School Chronicle*, vol. X, no. 3 (December 1906), pp. 56, 58.
14. Ibid., vol. XI, no. 3 (December 1907), p. 62.
15. Board of Education, inspection report, Bristol Grammar School, February 22–25, 1910 (Board of Education papers, ED.109/1600).
16. Ibid., October 26–29, 1915 (Board of Education papers, ED.109/1601).
17. Board of Education inspection reports 1910, 1915 (Board of Education papers, ED.109/1600, 1601).
18. Board of Education inspection report, BGS, 1910 (Board of Education papers, ED.109/1600).
19. Ibid., 1915 (Board of Education papers, ED.109/1601).
20. Meeting of BGS Board of Governors, December 14, 1906 (BGS Governors Minute Book, Bristol Record Office).
21. Cyril Norwood, annual report, September 1907 (Bristol Grammar School papers, Bristol Record Office).
22. Ibid.
23. Cyril Norwood, letter to P.J. Worsley, November 26, 1910 (BGS Governors Minute Book, Bristol Record Office).
24. Cyril Norwood, letter to Bristol education committee, June 8, 1911 (Meeting of BGS Board of Governors, June 9, 1911, Bristol Record Office).
25. Board of Education, inspection report, 1905 (Board of Education papers, ED.109/1599).
26. Board of Education, inspection report, 1910 (Board of Education papers, ED.1091/1600).
27. Hill, *The History of BGS*, p. 171.
28. Special Meeting of BGS Governors, July 26, 1907 (BGS Governors Minute Book, Bristol Record Office).

29. Board of Education, inspectors' report, Bristol Grammar School, 1915 (Board of Education papers, ED.109/1601).
30. *Bristol Grammar School Chronicle*, vol. XII, no. 6 (December 1911), p. 262.
31. Cyril Norwood, memorandum, " 'Proctor Baker' scholarship," March 10, 1911; Meeting of BGS Board of Governors, April 3, 1911 (Bristol Grammar School papers, Bristol Record Office).
32. Hill, *History of Bristol Grammar School*, pp. 131–32.
33. Meeting of BGS Board of Governors, June 9, 1916, report by Cyril Norwood on interview at the Board of Education on June 5, 1916 on "Peloquins and Free Places" (Bristol Grammar School papers, Bristol Record Office).
34. Cyril Norwood, memorandum, " 'Proctor Baker' scholarship," March 10, 1911; meeting of BGS Board of Governors, April 3, 1911 (Bristol Grammar School papers, Bristol Record Office).
35. Ibid.
36. Ibid.
37. Cyril Norwood, Report on the year 1912–1913 (Bristol Grammar School papers, Bristol Record Office).
38. Ibid.
39. Board of Education, inspection report, Bristol Grammar School, 1915 (Board of Education papers, ED.109/1601).
40. Ibid.
41. Bristol Grammar School, Order Book for Masters, curriculum and syllabus, Notes by Cyril Norwood, December 1, 1909 (Bristol Grammar School papers, BGS).
42. Board of Education, inspection report, Bristol Grammar School, 1910 (Board of Education papers, ED.109/1600).
43. Cyril Norwood, Report on External examination of the Grammar School; Meeting of BGS board of governors, June 11, 1909 (BGS Governors Minute Book, Bristol Record Office).
44. Board of Education, "Teaching of Latin in Secondary Schools" (Circular 574), October 10, 1907, p. 1.
45. Board of Education, inspection report, Bristol Grammar School, 1915 (Board of Education papers, ED.109/1601).
46. *Bristol Grammar School Chronicle*, vol. XI, no. 3 (December 1907), p. 63.
47. Board of Education, "The Teaching of English in Secondary Schools" (Circular 753), 1910, p. 1.
48. Ibid., p. 6.
49. Cyril Norwood, memorandum, "English—Syllabus 1913" (Bristol Grammar School, order book for masters, Bristol Grammar School papers).
50. Ibid; also Cyril Norwood, "English" (Bristol Grammar School, order book for masters, Bristol Grammar School papers).
51. Ibid.
52. Cyril Norwood and A.H. Hope (eds.), *The Higher Education of Boys in England* (London, John Murray, 1909), p. 341.
53. Board of Education, inspection report, Bristol Grammar School, 1915 (Board of Education papers, ED.109/1601).
54. Board of Education, "Teaching of History in Secondary Schools" (Circular 599), 1908, p. 1.
55. Ibid., p. 4.
56. Ibid., p. 5.
57. Cyril Norwood, "History Syllabus" (Bristol Grammar School, order book for masters, Bristol Grammar School papers).
58. Ibid.

59. Norwood and Hope (eds.), *Higher Education of Boys in England*, pp. 415–16.
60. Board of Education, inspection report, Bristol Grammar School, 1910 (Board of Education papers, ED.109/1600).
61. Ibid., 1915 (Board of Education papers, ED.109/1601).
62. *Bristol Grammar School Chronicle*, vol. XIII, no. 3 (December 1913), p. 142.
63. Ibid., vol. XIII, no. 5 (July 1914), p. 245.
64. *The Bristolian and Clifton Social World*, vol. III, no. 4 (May 1914), "Our Opinion of the YMCA," p. 78.
65. Cyril Norwood, annual report, September 1907 (Bristol Grammar School papers, Bristol Records Office).
66. Ibid.
67. *Bristol Grammar School Chronicle*, vol. XI, no. 3 (December 1907), p. 74.
68. Board of Education, inspection report, Bristol Grammar School, 1915 (Board of Education papers, ED.109/1601).
69. *Bristol Grammar School Chronicle*, vol. XI, no. 2 (August 1907), p. 28.
70. Ibid., vol. XII, no. 6 (December 1911), p. 223.
71. Ibid., vol. XII, no. 9 (December 1912), "Old Boys' Notes," p. 386.
72. Ibid., vol. XIII, no. 6 (December 1914), p. 257.
73. Ibid., vol. XIII, p. 305.
74. Board of Education, inspection report, Bristol Grammar School, 1910 (Board of Education papers, ED.109/1600).
75. Ibid., 1915 (Board of Education papers, ED.109/1601).
76. Cyril Norwood, letter, "To the Governors of Bristol Grammar School," n.d. [1906] (Bristol Grammar School papers, Bristol Record Office).
77. *Bristol Grammar School Chronicle*, vol. XI, no. 2 (August 1907), p. 34.
78. Ibid., vol. XII, no. 2 (July 1910), p. 46.
79. Ibid., vol. XI, no. 9 (December 1909), p. 338.
80. Bristol Grammar School Literary and Debating Society, Rules and Minutes, Introduction (Bristol Grammar School papers, Bristol Grammar School).
81. Bristol Grammar School Literary and Debating Society, general meeting, November 4, 1909; general meeting, April 7, 1910.
82. Bristol Grammar School, "Carmen Bristoliense," 1911 (Bristol Grammar School papers, Bristol Grammar School).
83. Cyril Norwood, Head master's report on Bristol Grammar School for 1915–1916 and the first term of 1916–1917 (Bristol Grammar School annual reports, Bristol Record Office); Cyril Norwood, memorandum, "School Pictures" (Bristol Grammar School order book for masters, Bristol Grammar School).
84. Cyril Norwood, "Discipline," "Detention" (in Cyril Norwood, Order book for masters, Bristol Grammar School papers).
85. *Bristol Grammar School Chronicle*, vol. XIV, no. 3 (December 1916), p. 106.
86. Norwood and Hope (eds.), *Higher Education of Boys in England*, p. 9.
87. Ibid., p. 21.
88. Ibid., p. 24.
89. Ibid., p. 28.
90. Ibid., p. 67.
91. Ibid., p. 155.
92. Ibid., pp. 160–61.
93. Ibid., p. 176.
94. Ibid., p. 197.
95. Ibid., p. 200.

96. Norwood and Hope (eds.), *Higher Education of Boys in England*, p. 209.
97. Ibid., p. 221.
98. Ibid., p. 225.
99. Ibid., p. 226.
100. Ibid., p. 280.
101. Ibid., pp. 289–90.
102. Ibid., p. 290.
103. Ibid., p. 559.
104. Ibid., p. 561.
105. Ibid., p. 562.
106. Ibid.

Chapter 5   Holding the Line?

1. Labour Party, *Secondary Education for All* (Labour Party, London, 1922).
2. *Journal of Education*, vol. 39, no. 578 (September 1917), p. 557, report on Oxford summer meeting, August 2–14.
3. W.N. Bruce, note, December 21, 1917 (Board of Education papers, ED.12/246).
4. Herbert Fisher, note, December 23, 1917 (Board of Education papers, ED.12/246).
5. See, e.g., John White, "The End of the Compulsory Curriculum," in Paul H. Hirst (ed.), *The Curriculum: The Doris Lee Lectures* (Studies in Education, New Series 2, University of London Institute of Education, 1975), pp. 22–39; Martin Lawn, "The Spur and the Bridle: Changing the Mode of Curriculum Control," *Journal of Curriculum Studies*, 19/3 (1987), pp. 227–36; and R. Brooks, "Lord Eustace Percy and the Abolition of the Compulsory, Elementary Curriculum in 1926," *Contemporary Record*, 7/1 (1993), pp. 86–102.
6. Lord Eustace Percy to G.B. Hurst, April 30, 1923 (Board of Education papers, ED.12/452).
7. Board of Education, "Curriculum of Secondary Schools" (Circular 826, 1913).
8. Committee on the position of natural science in the educational system of Great Britain, *Natural Science in Education* (HMSO, London, 1918), p. 2.
9. Ibid., p. 8.
10. Committee on the position of modern languages in the educational system of Great Britain, *Modern Studies* (HMSO, London, 1917).
11. Ibid., p. 136.
12. Report of the committee appointed by the prime minister to enquire into the position of classics in the educational system of the United Kingdom (HMSO, London, 1921).
13. Ibid., p. 6.
14. Ibid., p. 43.
15. Ibid., p. 268.
16. E. Pelham, note, November 15, 1922 (Board of Education papers, National Archives, ED.12/214).
17. Ibid.
18. Ibid.
19. Board of Education, "Curricula of Secondary Schools in England" (Circular 1294, 1922), paragraph 5.
20. Ibid., paragraph 6.
21. E. Pelham, memorandum to inspectors SN447, December 27, 1923 (Board of Education papers, ED.12/213).

22. See also, e.g., John Roach, "Examinations and the Secondary Schools, 1900–1945," *History of Education*, 8/1 (1979), pp. 45–58.
23. District Inspector for East Central and West Central Divisions, note, October 30, 1924 (Board of Education papers, ED.12/215).
24. Birmingham and Coventry inspectorate, note, February 3, 1924 (Board of Education papers, ED.12/215).
25. F.R.G. Duckworth, note, January 4, 1924 (Board of Education papers, ED.12/215).
26. G.T. Hawkins, note, January 14, 1924 (Board of Education papers, ED.12/215).
27. J.J.R. Budge, note, January 18, 1924 (Board of Education papers, ED.12/215).
28. F.B. Stead, note, January 24, 1924 (Board of Education papers, ED.12/215).
29. B.L. Pearson, note, February 29, 1924 (Board of Education papers, ED.12/215).
30. Board of Education, *Report of the Consultative Committee on Differentiation of the Curriculum for Boys and Girls Respectively in Secondary Schools* (HMSO, London, 1923), p. xii.
31. Ibid., p. xiii.
32. Ibid., p. xiv.
33. Ibid., p. 42.
34. Ibid., p. 126.
35. Sara Burstall to G.B. Hurst, MP, April 25, 1923 (Board of Education papers, ED.12/452).
36. Ibid.
37. Eastern and Metropolitan Division, note, February 2, 1924 (Board of Education papers, ED.12/215).
38. B.L. Pearson, note, February 29, 1924 (Board of Education papers, ED.12/215).
39. F.B. Stead, note, January 24, 1924 (Board of Education papers, ED.12/215).
40. E. Pelham, memorandum to inspectors S.454, July 21, 1924, "Curricula in Secondary Schools in England" (Board of Education papers, ED.12/215).
41. Association of Head Mistresses, memorandum to Board of Education, n.d. [March 1927], "The First School Examination" (Board of Education papers, ED.24/1636).
42. "The First Examination," discussion of AHM proposals, November 16, 1927 (Board of Education papers, ED.24/1636).
43. Ibid.
44. Cyril Norwood, "Girls Should *not* be Educated like Boys," *Daily Mail*, October 9, 1928.
45. *Time and Tide*, May 18, 1928, p. 479.
46. *Bath and Wilts Chronicle and Herald*, June 13, 1931.
47. See, e.g., Simon, *The Politics of Educational Reform*.
48. Labour Party, *Secondary Education for All: A Policy for Labour* (Labour Party, London, 1922), p. 7.
49. Ibid., p. 11.
50. Ibid., p. 29.
51. Board of Education, *Report of the Consultative Committee on the Education of the Adolescent* (Hadow Report) (HMSO, London, 1926), p. xxi.
52. Ibid., pp. xxi–xxii.
53. Ibid., p. xxii.
54. Ibid., p. xxiii.
55. Cyril Norwood to Secretary, Board of Education, October 17, 1928 (Board of Education papers, ED.12/255). See also McCulloch, *Failing the Ordinary Child?*, pp. 40–42.
56. C. Norwood to F.B. Stead, November 7, 1927 (Board of Education papers, ED.12/255).
57. Cyril Norwood, memorandum, "The School Certificate" [n.d.; January 1928?] (Board of Education papers, ED.12/255).
58. Ibid.

59. F.B. Stead, note to Maurice Holmes, February 20, 1928 (Board of Education papers, ED.12/255).
60. Ibid.
61. Standing Committee of SSEC, March 10, 1928 (Board of Education papers, ED.12/255).
62. Maurice Holmes, note, March 20, 1928 (Board of Education papers, ED.12/255).
63. W.R. Richardson, note, March 26, 1928 (Board of Education papers, ED.12/255).
64. Ibid.
65. F.B. Stead, note, May 16, 1928 (Board of Education papers, ED.12/255).
66. Cyril Norwood, "Education: The Next Steps" (presidential address to Educational Science section, British Association for the Advancement of Science, Glasgow, September 1928), report of 96th meeting of the British Association, September 5–12, 1928, p. 200. See also for an abridged version of this speech, Cyril Norwood, "Education: The Next Steps," *Journal of Education*, October 1928, pp. 521–24.
67. Ibid., p. 201.
68. Ibid., p. 202.
69. Ibid., p. 203.
70. Ibid., p. 206.
71. Ibid., p. 207.
72. Ibid., p. 210.

Chapter 6   Marlborough and Harrow

1. A.G. Bradley, A.C. Champneys, J.W. Baines, *A History of Marlborough College during Fifty Years from its Foundation to the Recent Time* (London, John Murray, 1893), p. 56.
2. Ibid., p. 54.
3. Clarendon report, 1864, vol. I, p. 2.
4. Frank Fletcher, *After Many Days: A Schoolmaster's Memories* (London, Robert Hale and Company, 1937), p. 108.
5. Ibid., p. 109.
6. Ibid., p. 111.
7. Cyril Norwood, application for Headship of Marlborough College, April 1916 (Norwood papers, 230/3/4).
8. *The Marlburian*, vol. LII, no. 774, May 30, 1917, p. 67.
9. Ibid.
10. Ibid., p. 68.
11. Supplement to *The Marlburian*, vol. LVI, no. 820 (1921), report of Prize Day, June 24, 1921.
12. *The Marlburian*, war memorial supplement, May 23, 1925.
13. Ibid., vol. LVI, no. 817 (1921), "Sermon by the Master," pp. 32–33.
14. Ibid., vol. LVII, no. 828, June 1, 1922, editorial, "Swindon Boys' Camp."
15. Cyril Norwood, confidential report to Council, November 1917 (Marlborough College archive).
16. Ibid.
17. *The Marlburian*, supplement, Prize Day, June 24, 1921.
18. T.C. Worsley, *Flannelled Fool: A Slice of Life in the Thirties* (London, Alan Ross Ltd, 1966).
19. Ibid., pp. 38–39.

20. Ibid., pp. 39–40.
21. Ibid., p. 40.
22. John Betjeman, *Summoned by Bells* (London, John Murray, 1960), p. 66.
23. Ibid., p. 67.
24. Ibid., pp. 68–69.
25. See also Bevis Hillier, *Young Betjeman* (London, John Murray, 1988).
26. M.J. Hayward, unpublished autobiography (Marlborough College archive).
27. See Jon Stallworthy and *Louis MacNeice* (London, Faber and Faber, 1995), esp. chapter 6.
28. Louis MacNeice, *The Strings are False: An Unfinished Autobiography* (London, Faber and Faber, 1965), p. 81.
29. Ibid., p. 85.
30. Ibid., p. 94.
31. See Miranda Carter, *Anthony Blunt: His Lives* (London, Picador, 2003); and John Costello, *Mask of Treachery* (London, Collins, 1988).
32. Carter, *Anthony Blunt*, p. 18.
33. MacNeice, *The Strings are False*, p. 95.
34. Cyril Norwood, sermon, March 28, 1921, in *The Marlburian*, vol. LVI, no. 817, p. 33 (Marlborough College archive).
35. Ibid.
36. Cyril Norwood, "Marlborough education," in *Marlborough College 1843–1943* (Cambridge, Cambridge University Press, 1943), p. 42.
37. Ibid., p. 43.
38. *The Heretick*, editorial, no. 1, March 1924, p. 1 (Marlborough College archive).
39. *The Heretick*, article, "The Road," no. 1, March 1928, pp. 4–5 (Marlborough College archive).
40. Old Marlburian, notes on Marlborough College, c. 1967 (Marlborough College archive).
41. *The Heretick*, no. 2, June 1924, articles, "Art and morality," p. 10, and "Socialism," p. 12 (Marlborough College archive).
42. G.V.S. Bucknall, note, February 15, 1991 (Marlborough College archive).
43. Former Marlborough College pupil, letter to the author, March 16, 2000.
44. C.M. Bowra, *Memories, 1898–1939* (London, Weidenfeld and Nicolson, 1966), p. 165.
45. John Bowle, interview, 1976, quoted in Hillier, *Young Betjeman*, p. 93.
46. Former Marlborough College pupil, letter to the author, April 10, 2000.
47. *The Marlburian*, vol. LIX, no. 848, March 5, 1924, p. 23 (Marlborough College archive).
48. Ibid., vol. LIX, no. 853, March 31, 1924, pp. 34–35 (Marlborough College archive).
49. John Baines, unpublished autobiography (Marlborough College archive).
50. Former Marlborough College pupil, letter to the author, March 20, 2000.
51. Ibid., April 5, 2000.
52. Ibid., April 8, 2000.
53. MacNeice, *The Strings are False*, p. 81.
54. Worsley, *Flannelled Fool*, p. 40.
55. John Baines, unpublished autobiography (Marlborough College archive).
56. Christopher Bell, "Random memories of Marlborough in early 1920s" (Marlborough College archive).
57. M.J. Hayward, unpublished autobiography (Marlborough College archive).
58. *The Marlburian*, vol. LVI, no. 818, May 24, 1921, p. 70.
59. Board of Education inspection report on Marlborough College, 26–30 1924 (Board of Education papers, ED.109/6596).
60. *Country Life*, no. 1511, December 19, 1925.
61. *The Times*, obituary of Sir Cyril Norwood, March 14, 1956.

62. R.F., Cyril Norwood: a courageous leader, *The Times*, March 20, 1956; see also *The Times*, March 15, 1956.
63. *The Harrovian*, May 3, 1956, obituary of Sir Cyril Norwood.
64. Giles Playfair, *My Father's Son* (London, Geoffrey Bles), pp. 109, 113, 114.
65. Christopher Tyerman, *A History of Harrow School* (London, Oxford University Press, 2000), p. 512.
66. Cyril and Catherine Norwood, note, "For Staying, for Going" [n.d.; 1933] (Norwood papers, 230/5/13).
67. Jonathan Gathorne-Hardy, *The Public School Phenomenon, 597–1977* London, 1977), p. 302.
68. Tim Card, *Eton Renewed: A History From 1860 to the Present Day* (London, John Murray, 1994), p. 145.
69. Cyril Alington to Cyril Norwood, December 5, 1925 (Norwood papers, 230/4/7).
70. Tyerman, *A History of Harrow School*.
71. Board of Education, inspection report on Harrow School, October 12–16, 1931 (Board of Education papers, National Archives, ED.109/4198).
72. Ibid.
73. Ibid.
74. Anthony Part, *The Making of a Mandarin* (London, Andre Deutsch, 1990), p. 9.
75. Francis Pember to A.V. Hill, December 23, 1928 (Hill papers, Churchill College Cambridge).
76. Ibid., November 17, 1933 (Hill papers).
77. A.V. Hill to Francis Pember, December 14, 1933 (Hill papers).
78. Board of Education inspection report on Harrow School, October 12–16, 1931 (Board of Education papers, ED.109/4198).
79. Former Harrow pupil, letter to the author, July 27, 2000.
80. Ibid., June 12, 2000.
81. Ibid., July 27, 2000.
82. Ibid.
83. Ibid., July 25, 2000.
84. Ibid., July 2000.
85. Ibid., August 8, 2000.
86. Ibid., July 25, 2000.
87. Ibid., July 7, 2000.
88. Ibid., July 27, 2000.
89. Ibid., June 1, 2000.
90. Ibid., August 25, 2000.
91. Tyerman, *Harrow School*, p. 315.
92. *The Harrovian*, February 26, 1959, obituary, C.G.P.
93. Former Harrow pupil, letter to the author, August 29, 2000.
94. *The Times*, February 11, 1959, obituary, Mr. C.G. Pope.
95. Former Harrow pupils, letters to the author, July 5, 2000, June 1, 2000.
96. Ibid., August 3, 2000.
97. Ibid., May 31, 2000, June 1, 2000.
98. Ibid., August 3, 2000.
99. Ibid., July 24, 2000.
100. Ibid., July 27, 2000.
101. C. Norwood to C.G. Pope, July 29, 1929 (Norwood papers, 230/5/6).
102. Norwood to Pope, June 25, 1929 (Norwood papers, 230/5/6).
103. Ibid.
104. Percival Hardy (solicitor) to Cyril Norwood, July 22, 1929 (Norwood papers, 230/5/6).

105. Norwood to Hardy, June 27, 1929; Hardy to Norwood, July 22, 1929; Francis Pember to Norwood, June 26, 1929; Norwood to C.G. Pope, July 29, 1929 (Norwood Papers, 230/5/6).
106. Pope to Norwood, June 30, 1929 (Norwood papers, 230/5/6).
107. Norwood to Pope, July 13, 1929 (Norwood papers, 230/5/6).
108. Pope to Pember, July 19, 1929 (Norwood papers, 230/5/6).
109. Pope to Norwood, July 29, 1929 (Norwood papers, 230/5/6).
110. Norwood to Pope, July 25, 1929 (Norwood papers, 230/5/6).
111. Cyril Norwood, memorandum to Harrow board of governors, August 1929 (Harrow School archive).
112. Harrow School, board of governors, meeting, November 13, 1929 (Harrow School archive).
113. Pope to Norwood, November 15, 1929; Norwood to Pope, November 15, 1929 (Norwood papers, 230/5/6).
114. Ibid., January 19, 1930 (Norwood papers, 230/5/6).
115. *The Harrovian*, February 26, 1959, obituary, C.G.P.
116. Francis Pember to Catherine Norwood, March 29, 1930 (Norwood papers, 230/5/6).
117. Ibid., April 3, 1930 (Norwood papers, 230/5/6).
118. J.R. de S. Honey, *Tom Brown's Universe: The Development of the Public School in the 19th Century* (London, Millington, 1977), pp. 12–13.
119. Cyril Norwood to Harrow pupil, August 4, 1931 (private letter. I am most grateful to this former Harrow pupil for making this letter available to me.)
120. Cyril and Catherine Norwood, "For Staying, for Going" [n.d.; 1933] (Norwood Papers, 230/5/13).
121. Cyril Norwood to William Costin, November 1, 1933 (St. John's College Oxford archive, Munim L111 A137).
122. Ibid.

Chapter 7 The English Tradition of Education

1. See also Gary McCulloch and Colin McCaig, "Reinventing the Past: The Case of the English Tradition of Education," *British Journal of Educational Studies*, 50/2 (2002), pp. 238–53.
2. Mills, *The Sociological Imagination*.
3. Cyril Norwood, "The Public Schools," *Journal of Education and School World* (May 1926), p. 317.
4. Ibid., p. 318.
5. Ibid.
6. Cyril Norwood, "Public Schools and Social Service," *The Spectator*, November 13, 1926, p. 847.
7. Ibid., p. 848.
8. Ibid.
9. Cyril Norwood, "The Boys' Boarding School," in J. Dover Wilson (ed.), *The Schools of England: A Study in Renaissance* (London, Sidgwick and Jackson, 1928), p. 131.
10. Ibid., p. 134.
11. Ibid., p. 135.
12. Ibid., p. 136.

13. Sir Lewis Selby-Bigge, *The Board of Education* (London, Putnam's, 1927), Preface.
14. Cyril Norwood, *The English Educational System* (London, Ernest Benn Ltd, 1927), p. 9.
15. Ibid., pp. 10–11.
16. Ibid., p. 29.
17. Ibid., p. 30.
18. Ibid.
19. Ibid., p. 38.
20. Ibid., p. 48.
21. Ibid., p. 58.
22. Ibid., pp. 78–79.
23. Ibid., p. 78.
24. Cyril Norwood, *The English Tradition of Education* (London, John Murray, 1929), p. v.
25. Ibid., pp. 3–4.
26. Ibid., p. 4.
27. Ibid., p. 6.
28. Ibid., p. 9.
29. Ibid., p. 17.
30. Ibid., p. 19.
31. Ibid., p. 20.
32. Ibid., p. 57.
33. Ibid., p. 109.
34. Ibid., p. 122.
35. Ibid., p. 8.
36. See Herbert Butterfield, *The Whig Interpretation of History* (1931/1973) (London, Penguin, 1973), p. 9.
37. Norwood, *English Tradition of Education*, pp. 154–55.
38. Ibid., pp. 166–67.
39. Ibid., p. 171.
40. Ibid., p. 178.
41. Ibid., pp. 183–84.
42. Ibid., p. 184.
43. Ibid., p. 185.
44. Ibid., p. 188.
45. Ibid., p. 208.
46. Ibid.
47. Ibid., p. 219.
48. Bertrand Russell, *On Education* (London, Allen and Unwin, 1926).
49. Ibid., pp. 239–40.
50. Ibid., p. 249.
51. Ibid., p. 300.
52. Ibid., p. 335.
53. Ronald Gurner to Cyril Norwood, October 2, 1929 (Norwood papers, 230/5/6).
54. T.F. Coade to Cyril Norwood, October 19, 1929 (Norwood papers, 230/5/6).
55. Dr G.F. Morton to Cyril Norwood, October 23, 1929 (Norwood papers, 230/5/6).
56. Lt. Col. John Murray to Cyril Norwood, November 1, 1929 (Norwood papers, 230/5/6).
57. Brigadier G.S., letter to Cyril Norwood, October 2, 1929 (Norwood papers, 230/5/6).
58. C.W. Valentine, "The English Tradition in Education," *Forum of Education*, 8/1 (February 1930), pp. 55–57.
59. Lord Gorell, "The Tide of English Education," *Quarterly Review*, October 1929, p. 330.
60. Frederick J. Mathias, "Vital Utterances on Education," *Western Mail*, November 7, 1929.

61. *The Tablet Literary Supplement*, "A Headmaster Speaks out," December 7, 1929; *New Statesman*, "Our Public Schools," November 25, 1929; Johannesburg *Sunday Times*, review, November 10, 1929; *Otago Daily News*, editorial, "The Real Thing in Education," November 29, 1930.
62. *Times Literary Supplement*, "Tradition and Education," October 10, 1929.
63. Ibid.
64. *The Times*, "Dr Norwood on Education: The English Tradition," October 12, 1929.
65. Ibid.
66. Sir Michael Sadler, "The English Public School Spirit," *The Observer*, October 20, 1929.
67. H.C. Dent, "An Educational Survey. II. The Aim: An Educated Democracy," *The Nineteenth Century and After*, DCXXXV (January 1930), p. 11.
68. Ibid., p. 14.
69. Ibid., p. 16.
70. Kingsley Martin, "Public Schools," *Time and Tide*, October 25, 1929, p. 1278.
71. Ibid.

Chapter 8   The New World of Education

1. Cyril Norwood to Catherine Norwood, January 6, 1935 (Norwood papers, 230/6/3).
2. See, e.g., T.F. Coade (ed.), *Harrow Lectures on Education* (Cambridge, Cambridge University Press, Cambridge, 1931); E.D. Laborde (ed.), *Education of To-Day* (Cambridge, Cambridge University Press, 1935); and E.D. Laborde (ed.), *Problems in Modern Education* (Cambridge, Cambridge University Press, 1939).
3. Cyril Norwood to Catherine Norwood, April 13, 1927 (Norwood papers, University of Sheffield).
4. Ibid., March 28, 1930 (Norwood papers, 230/5/8).
5. Ibid., April 23, 1930 (Norwood papers, 230/5/8).
6. See, e.g., Simon, *The Politics of Educational Reform*, chapter 4.
7. Cyril Norwood, "The Education and Outlook of English Youth," address given to the Canadian Club of Montreal at the Windsor Hotel Montreal, April 11, 1930 (Norwood papers, 230/5/8).
8. Cyril Norwood, "Unity and Purpose in Education," in T.F. Coade (ed.), *Harrow Lectures on Education* (Cambridge, Cambridge University Press, 1931), p. 3.
9. Ibid., p. 7.
10. Ibid., p. 9.
11. Cyril Norwood, *Religion and Education* (Teaching Christ papers no. IX, St. Christopher Press, Society for Promoting Christian Knowledge, 1932), p. 6.
12. Ibid., p. 7.
13. Ibid., p. 16.
14. Cyril Norwood, "Scylla and Charybdis, or Laissez-Faire and Paternal Government" (9th Shaftesbury lecture, delivered on Monday, May 2, 1932, Kingsgate Chapel, Shaftesbury Society, London, 1932), p. 5.
15. Ibid., p. 10.
16. Ibid.
17. Ibid., p. 13.
18. Ibid.

19. Matthew Grimley, *Citizenship, Community and the Church of England: Liberal Anglican Theories of the State between the Wars* (Oxford, Clarendon Press, 2004), pp. 6–7.
20. Cyril Norwood, "Education and Citizenship," presidential address to the Science Masters Association, December 29, 1931, p. 1.
21. Ibid., p. 9.
22. On the influence of eugenics in social and political thought between the Wars, see, e.g., Elazar Barkan, *The Retreat of Scientific Racism: Changing Concepts of Race in Britain and the United States between the World Wars* (Cambridge, Cambridge University Press, 1992).
23. Norwood, "Education and Citizenship," p. 16.
24. Ibid., p. 19.
25. Rt Hon Edward Lyttelton to Lord Irwin, May 17, 1933 (Board of Education papers, National Archives, ED.12/452).
26. Ibid. Whitehead's influential book *The Aims of Education and other Essays* had been published in 1932 (London, Williams and Norgate).
27. Ibid.
28. F.B. Stead, note, May 29, 1933 (Board of Education papers, ED.12/452).
29. Board of Education, *Report of the Consultative Committee on Secondary Education with Special Reference to Grammar Schools and Technical High Schools* (London, HMSO, 1939), p. iv, Terms of reference.
30. Cyril Norwood, memorandum to Spens Committee on secondary education, n.d. [1933] (Board of Education papers, ED.10/151).
31. Cyril Norwood, oral evidence to Spens Committee, November 24, 1933 (Board of Education papers, ED.10/151).
32. Will Spens to Ernest Simon, June 30, 1934 (Simon papers, Manchester Central Library, M11/14/14).
33. Sir Ernest Simon to T.F. Coade, August 25, 1934 (Simon papers, M11/14/14).
34. Sir Ernest Simon to Eva Hubback, December 23, 1936 (Simon papers, M11/14/15).
35. Ibid., December 24, 1936 (Simon papers, M11/14/15).
36. Sir Ernest Simon to Spencer Leeson, December 30, 1936 (Simon papers, M11/14/15).
37. Spencer Leeson to Sir Ernest Simon, n.d. [January 1937] (Simon papers, M11/14/15).
38. Sir Ernest Simon to Spencer Leeson, January 14, 1937 (Simon papers, M11/14/15).
39. Spencer Leeson to Sir Ernest Simon, n.d. [January 1937] (Simon papers, M11/14/15).
40. Cyril Norwood, *The Curriculum in Secondary Schools* (London, Association for Education in Citizenship, 1937), p. 5.
41. Ibid., pp. 5–6.
42. R.F. Young to Sir Ernest Simon, March 31, 1937 (Simon papers, M11/14/15).
43. Hugh Lyon to Sir Ernest Simon, January 1, 1937 (Simon papers, M11/14/15).
44. Sir Ernest Simon to Eva Hubback, June 9, 1937 (Simon papers, M11/14/15).
45. Sir Ernest Simon to Will Spens, February 17, 1938 (Simon papers, M11/14/15).
46. Ibid., March 9, 1938 (Simon papers, M11/14/15). See also McCulloch, *Philosophers and Kings*, chapter 3.
47. Spens report, p. xxxi.
48. Ibid., p. xxxvii.
49. Ibid., pp. xxvii–xxviii.
50. Ibid., pp. 145–46.
51. Ibid., p. 142.
52. Ibid., pp. 142–43.
53. Spens report, Appendix IV, pp. 429–38.
54. Ibid., p. 219.

55. Ibid., p. 236.
56. Ibid., pp. 244–45.
57. See, e.g., A.E. Campbell (ed.), *Modern Trends in Education: The Proceedings of the New Education Fellowship Conference held in New Zealand in July 1937* (Wellington, Whitcombe and Tombs, 1938), pp. 496–97.
58. Peter Fraser to K.S. Cunningham, October 27, 1937 (New Zealand Education Department papers, National Archives, Wellington, file 4/10/26). See also, e.g., Jane Abbiss, "The 'New Education Fellowship' in New Zealand: Its Activity and Influence in the 1930s and 1940s," *New Zealand Journal of Educational Studies*, 33/1 (1998), pp. 81–93.
59. C.E. Beeby to Peter Fraser, October 27, 1937 (New Zealand Education Department papers, 4/10/26).
60. Cyril Norwood, diary, June 9, 1937 (Norwood papers, 230/6/6); Catherine Norwood, diary, June 10, 1937 (Norwood papers, 230/10/8), The following references are from these diary sources also.
61. Catherine Norwood, diary, June 18, 1937; Cyril Norwood, diary, June 10, 1937.
62. Cyril Norwood, diary, June 12, 1937.
63. Ibid., June 17, 1937.
64. Ibid., July 6, 1937; Catherine Norwood, diary, July 6, 1937.
65. Catherine Norwood, diary, June 19, 1937; July 8, 1937.
66. Ibid., July 14, 1937.
67. Ibid., July 13, 1937.
68. Ibid., July 17, 1937.
69. Ibid., August 9, 1937.
70. Ibid., September 6, 1937.
71. Cyril Norwood, diary, August 5, 1937.
72. Cyril Norwood, "The Educational, Social, and International Relevance of Christianity in the Modern World," in E.D. Laborde (ed.), *Problems in Modern Education* (Cambridge, Cambridge University Press, 1939), p. 10.
73. *Sydney Mail*, report, "Education—A Conference that May Lead to Reform," August 18, 1937, p. 10.
74. I.L. Kandel, "School and Society," in Campbell (ed.), *Modern Trends in Education*, pp. 1–12.
75. Ibid., p. 3.
76. Cyril Norwood, "Christianity and the World Crisis," "The New Conception of Physical Education," "Science and its Place in a General Education," "Music and its Place in Education," in Campbell (ed.), *Modern Trends in Education*, pp. 50–55, 184–87, 206–13, 213–15.
77. Canford School, meeting of Council, September 23, 1927 (Canford minute book, Allied Schools, Banbury).
78. Ibid., October 28, 1927.
79. *Bath and Wilts Chronicle and Herald*, November 6, 1934, report, "Secrets of a School: Monkton Combe Vicar's Achievements," November 6, 1934; Central Committee of the Allied Schools, October 26, 1934 (Allied Schools, Banbury).
80. Cyril Norwood, "Sons of the Poor and Schools of the Rich," *The Nineteenth Century and After*, no. 700, vol. CXVII (June 1935), p. 693.
81. Ibid., p. 694.
82. Ibid., p. 695.
83. Ibid., p. 696.
84. Cyril Norwood to P.R.G. Duckworth, October 19, 1938 (Board of Education papers, ED.136/129).

85. Note of interview with Cyril Norwood by G.G. Williams and F.R.G. Duckworth, October 24, 1938 (Board of Education papers, ED.136/129).
86. Ibid.
87. Fred Clarke, *Education and Social Change* (London, Sheldon Press, 1940).
88. Cyril Norwood to G.G. Williams, December 3, 1939 (Board of Education papers, ED.12/518).
89. Ibid.
90. Cyril Norwood, "The Crisis in Education—I," *The Spectator*, February 9, 1940, p. 176.
91. Cyril Norwood, "The Crisis in Education—II," *The Spectator*, February 16, 1940, p. 207.
92. Lord Hugh Cecil to Lord de la Warr, March 21, 1940 (Board of Education papers, ED.136/129).
93. Lord de la Warr to Lord Hugh Cecil, March 29, 1940 (Board of Education papers, ED.136/129).
94. *Times Educational Supplement*, leading article, "Public Schools," March 9, 1940.
95. T.C. Worsley, *Barbarians and Philistines: Democracy and the Public Schools* (London, Robert Hale Ltd, 1940), p. 131.
96. Ibid., p. 136. Worsley incorrectly dated Norwood's *English Tradition* as having been published in 1932.
97. Ibid., p. 263.
98. Cyril Norwood, "Democracy and the Public Schools," *Journal of Education*, November 1940, p. 481.
99. Ibid., p. 482.

Chapter 9   The Norwood Report and Secondary Education

1. Cyril and Catherine Norwood, note, "For Staying, for Going," n.d. [1933] (Norwood papers, 230/5/13).
2. W.C. Costin to Cyril Norwood, October 31, 1933 (St. John's College Oxford papers, Munim LIII).
3. Ibid.
4. F.B. Malim, diary, November 28, 1937 (Malim diary, private).
5. Costin to Norwood, October 31, 1933.
6. Malim, diary, November 28, 1937.
7. University of Oxford Hebdomadal Council, June 10, 1935, HC1/1/161; June 8, 1936, HC Acts (Hebdomadal council minutes, University of Oxford, Bodleian Library Oxford).
8. University of Oxford Hebdomadal Council, October 25, 1937, HC1/1/168 (Hebdomadal council minutes, University of Oxford).
9. *Oxford*, 6/1 (1939), pp. 24–25.
10. Hebdomadal council, January 29, 1940 (HC minutes, HC1/1/175).
11. George Smith, report, "Change of Title of Department for the Training of Teachers," 1936 (HC minutes, HC1/1/165).
12. Hebdomadal council, February 20, 1939 (HC minutes, HC1/1/172).
13. Cyril Norwood, "Oxford and Physical Education," *The Oxford Magazine*, June 2, 1938, p. 701.
14. Nevill Coghill, "Gym-Lido," *The Oxford Magazine*, June 16, 1938, pp. 764–65.
15. *The Modern Churchman*, vol. 46, no. 1 (March 1956), pp. 1–4. See also Cyril Norwood, "Christian Modernism and Education," *The Modern Churchman*, vol. 38 (1948), pp. 226–33.

16. Frank Hopkinson, letter to the author, November 5, 1999.
17. See Paul Addison, "Oxford and the Second World War," in Brian Harrison (ed.), *The History of the University of Oxford*, vol. VIII, The Twentieth Century (Oxford, Clarendon Press, 1994), pp. 167–88.
18. Andrew Motion, *Philip Larkin: A Writer's Life* (London, Faber and Faber, 1993), p. 54.
19. St. John's College Oxford, junior common room suggestion book, June 1939 (St. John's College Oxford papers, Muniments Room).
20. Suggestions book, St. John's Oxford, 1941 (St. John's College Oxford papers).
21. Suggestions book, St. John's Oxford, 1937 (St. John's College Oxford papers).
22. Spens Report (1939), Introduction.
23. Ibid., p. 293.
24. Ibid., chapter 9.
25. Lady Simon to Will Spens, May 21, 1938 (Lady Simon papers, Manchester Central Library, M14/2/2/3).
26. R.S. Wood, memorandum, July 30, 1937 (Board of Education papers, ED.10/273).
27. G.G. Williams, "Note on the Spens Report: Scarborough, 16/6/39" (Board of Education papers, ED.136/131).
28. Maurice Holmes, note to president of Board of Education, July 5, 1939 (Board of Education papers, ED.136/131).
29. SSEC Notes for Chairman, Agenda, 102nd meeting, December 14, 1940 (Board of Education papers, ED.12/532).
30. Herbert Ramsbotham to Cyril Norwood, January 3, 1941; Norwood to Ramsbotham, January 8, 1941 (Board of Education papers, ED.12/478).
31. Cyril Norwood, "Some Aspects of Educational Reconstruction," *The Fortnightly*, February 1941, pp.105–13.
32. Dorothy Brock to Cyril Norwood, February 8, 1941 (Norwood papers, 230/6/17).
33. See R.A. Butler, *The Art of the Possible* (London, Penguin, 1973), p. 10.
34. R.A. Butler, note on meeting with Cyril Norwood, July 31, 1941 (Board of Education papers, ED.12/478).
35. Cyril Norwood to G.G. Williams, August 13, 1941 (Board of Education papers, ED.12/478).
36. Norwood Report (1943), p. iv.
37. Cyril Norwood to R.A. Butler, September 20, 1941 (Board of Education papers, ED.12/478).
38. R.A. Butler, note on meeting with Cyril Norwood, November 27, 1941 (Board of Education papers, ED.12/478).
39. Ibid.
40. Ibid.
41. Minutes of first meeting of Norwood committee, October 18, 1941 (Board of Education papers, ED.136/581).
42. G.G. Williams, "Note on Future Policy in Secondary Education," December 4, 1941 (Incorporated Association of Assistant Masters in Secondary Schools papers, E1/2 file 1 (Institute of Education London).
43. Cyril Norwood, "Detailed Agenda for Meetings," n.d. [December 1941, for meetings January 5–7, 1942] (IAAM papers, E1/2 file 1).
44. Ibid.
45. Cyril Norwood, "A Note on the Grammar School," n.d. [prepared for Norwood committee meeting, January 5–7, 1942] (Board of Education papers, ED.136/681).
46. Ibid.
47. M.G. Holmes, memorandum to R.A. Butler, October 20, 1941 (Board of Education papers, ED.12/478).

48. G.G. Williams to Cyril Norwood, November 26, 1941 (Board of Education papers, ED.136/681).
49. R.A. Butler, note, March 20, 1942 (Board of Education papers, ED.136/131).
50. See also Gary McCulloch, " 'Spens v. Norwood': Contesting the Educational State?," *History of Education*, 22/2 (1993), pp. 163–80.
51. Cyril Norwood to R.A. Butler, June 6, 1942 (Board of Education papers, ED.136/681).
52. Cyril Norwood to G.G. Williams, June 6, 1942 (Board of Education papers, ED.12/478).
53. R.A. Butler, note, "The Common School," January 23, 1942 (Board of Education papers, ED.136/294).
54. Ibid.
55. M.G. Holmes, note, January 26, 1942 (Board of Education papers, ED.136/294).
56. R.A. Butler, note, May 28, 1942 (Board of Education papers, ED.136/681).
57. G.G. Williams, note to R.A. Butler and M.G. Holmes, December 23, 1941 (Board of Education papers, ED.12/478).
58. R.A. Butler, note, January 14, 1942 (Board of Education papers, ED.136/681).
59. Ibid., May 28, 1942.
60. McNair committee, oral evidence by Sir Cyril Norwood, September 29, 1942 (McNair committee papers, University of Liverpool).
61. Ibid.
62. Norwood committee, minutes of meeting, September 1–4, 1942; background paper 28, "Notes on the Curriculum," September 1942 (IAAM papers, E1/1 file 3).
63. Cyril Norwood to R.A. Butler, September 6, 1942 (Board of Education papers, ED.136/681).
64. Cyril Norwood to George Turner, September 27, 1942 (St. John's College papers, Muniments Room).
65. Cyril Norwood to R.A. Butler, September 6, 1942 (Board of Education papers, ED.136/681).
66. R.A. Butler, note, May 21, 1943 (Board of Education papers, ED.136/681).
67. Cyril Norwood to R.A. Butler, June 23, 1943 (Board of Education papers, ED.136/681).
68. R.A. Butler, note, May 21, 1943 (Board of Education papers, ED.136/681).
69. R.A. Butler, note to M.G. Holmes and G.G. Williams, June 6, 1943 (Board of Education papers, ED.136/681).
70. Norwood Report (1943), p. 2.
71. Ibid., p. 4.
72. Ibid., p. 3.
73. Board of Education, *The Public Schools and the General Educational System* (HMSO, London, 1944) (Fleming Report).
74. Norwood Report (1943), p. 17.
75. Ibid., p. 18.
76. Ibid., pp. 14, 24.
77. Ibid., pp. 4–5.
78. Cyril Norwood to G.G. Williams, June 10, 1943 (Board of Education papers, ED.12/478).
79. Final minutes of the Norwood committee (unofficial), June 23, 1943 (Board of Education papers, ED.12/479).
80. J.A. Lauwerys, "Curriculum and the Norwood Report," *Journal of Education*, vol. 76, April 1944, pp. 164–68.
81. Cyril Norwood, "The Norwood Report—An Unrepentant Statement," *Journal of Education*, vol. 76, May 1944, pp. 215–18.
82. Frederick S. Boas (President) and Arundel Esdaile (Chairman of Committees), The English Association, letter to *The Times*, August 19, 1944.

83. Cyril Norwood, letter to *The Times*, August 22, 1944.
84. Cyril Norwood to G.G. Williams, November 10, 1943 (Board of Education papers, ED.12/480).
85. J.A. Petch, *Fifty Years of Examining: The Joint Matriculation Board, 1903–1953* (London, Harrap, 1953), p. 165.
86. Cyril Norwood to R.A. Butler, November 19, 1943 (Board of Education papers, ED.12/480).
87. SSEC, minutes of 103rd meeting, November 19, 1943 (Board of Education papers, ED.12/480).
88. Cyril Norwood to G.G. Williams, May 11, 1944 (Board of Education papers, ED.12/480).
89. R.H. Barrow to Cyril Norwood, December 22, 1947 (Norwood papers, University of Sheffield).
90. Cyril Norwood, "Democracy, Freedom and Education," *The Fortnightly*, April 1943, p. 244.
91. Cyril Norwood to George Turner, September 27, 1942 (St. John's College Oxford papers).
92. Fred Clarke, note, August 21, 1943 (Clarke papers, Institute of Education London, file 36).
93. See Gary McCulloch, "Local Education Authorities and the Organisation of Secondary Education, 1943–1950," *Oxford Review of Education*, 28/2–3 (2002), pp. 235–46.
94. Cyril Norwood to G.G. Williams, May 11, 1944 (Board of Education papers, ED.12/480).
95. R.D. Heaton, note, January 3, 1944 (Board of Education papers, ED.136/592).
96. R.A. Butler, note, August 16, 1944 (Board of Education papers, ED.136/592).
97. G.G. Williams, note to M.G. Holmes, August 10, 1945 (Ministry of Education papers, ED.147/133).
98. M.G. Holmes, note to Ellen Wilkinson, August 11, 1945 (Ministry of Education papers, ED.147/133).
99. G.G. Williams, note, January 30, 1946 (Ministry of Education papers, ED.147/133).

Chapter 10   Conclusions: The Ideal of Secondary Education

1. Banks, *Parity and Prestige in Secondary Education*; Simon, *The Politics of Educational Reform*.
2. For full discussions of these major figures see, e.g., Douglas J. Simpson, *John Dewey and the Art of Teaching: Toward Reflection and Imaginative Practice* (London, Sage, 2005); Geoffrey Walford and W.S.F. Pickering (eds.), *Durkheim and Modern Education* (London, Routledge, 1998); J.H. Higginson (ed.), *Selections from Michael Sadler: Studies in World Citizenship* (Liverpool, Dejallot Meyerre, 1979). Ross Terrill, *R.H. Tawney and His Times: Socialism as Fellowship* (London, Deutsch, 1974); Leslie Hearnshaw, *Cyril Burt, Psychologist* (London, Hodder and Stoughton, 1979); and Noeline Alcorn, *To the Fullest Extent of His Powers: C.E. Beeby's Life in Education* (Wellington, Victoria University Press, 1999).
3. See, e.g., Barry M. Franklin and Gary McCulloch, "Partnerships in a 'Cold Climate': The Case of Britain," in B. Franklin, M. Bloch, and T. Popkewitz (eds.), *Educational Partnerships and the State: The Paradoxes of Governing Schools, Children, and Families* (New York, Palgrave Macmillan, 2003), pp. 83–107; Gary McCulloch, "From Incorporation to Privatisation: Public and Private Secondary Education in Twentieth-Century England," in R. Aldrich (ed.), *Public or Private Education? Lessons from History* (London, Woburn,

2004), pp. 53–72; and Geoffrey Walford (ed.), *British Private Schools: Research on Policy and Practice* (London, Woburn, 2003).
4. Department for Education and Skills, *Higher Standards, Better Schools for All: More Choice for Parents and Pupils* (London, Stationery Office, 2005). Alison Shepherd, "Blair Only Cares about Pushy Middle Classes," *Times Educational Supplement*, October 28, 2005, p. 22; Peter Wilby, "The Richest are White Paper Winners," *TES*, November 4, 2005, p. 23.

Bibliography

1. Archival Sources

Allied Schools papers, Banbury
Bristol Grammar School papers, Bristol Grammar School
Bristol Grammar School papers, Bristol Record Office
Fred Clarke papers, Institute of Education, University of London
Education department/Board of Education papers, National Archives, Kew
A.V. Hill papers, Churchill College, Cambridge
Incorporated Association of Assistant Masters in Secondary Schools, papers, Institute of Education, University of London
Leeds Grammar School papers, Leeds Grammar School
McNair committee papers, University of Liverpool
F.B. Malim papers, private
Marlborough College papers, Marlborough College
New Zealand Education Department papers, National Archives, Wellington
Cyril Norwood papers, University of Sheffield
St John's College Oxford papers, St John's College Oxford
Lord (Ernest) Simon papers, Manchester Central Library
Lady (Shena) Simon papers, Manchester Central Library
University of Oxford papers, Bodleian Library Oxford

2. Correspondence and Interviews

Correspondence with former pupils at Harrow School
Correspondence with former pupils at Marlborough College

3. School and University Magazines

Bristol Grammar School Chronicle (Bristol Grammar School magazine)
Harrovian (Harrow School magazine)
The Heretick (Marlborough College magazine)
The Leodiensian (Leeds Grammar School magazine)

The Marlburian (Marlborough College magazine)
Oxford (University of Oxford)
The Oxford Magazine (University of Oxford)
The Taylorian (Merchant Taylors School magazine)

4. Newspapers and Periodicals (selected)

Country Life
The Fortnightly
Journal of Education
The Modern Churchman
The Nineteenth Century and After
The Spectator
Time and Tide
The Times
The Times Educational Supplement
The Times Literary Supplement

5. Reports

Board of Education (1923) *Report of the Consultative Committee on Differentiation of the Curriculum for Boys and Girls Respectively in Secondary Schools*, London, HMSO
Board of Education (1926) *Report of the Consultative Committee on the Education of the Adolescent*, London, HMSO (Hadow Report)
Board of Education (1939) *Report of the Consultative Committee on Secondary Education with Special Reference to Grammar Schools and Technical High Schools*, London, HMSO (Spens Report)
Board of Education (1943) *Curriculum and Examinations in Secondary Schools*, London, HMSO (Norwood Report)
Board of Education (1944) *The Public Schools and the General Educational System*, London, HMSO
Committee on the position of modern languages in the educational system of Great Britain (1917) *Modern Studies*, London, HMSO
Committee on the position of natural science in the educational system of Great Britain (1918) *Natural Science in Education*, London, HMSO
Crockford's Clerical Directory (annual)
Department for Education and Skills (2005) *Higher Standards, Better Schools for All: More Choice for Parents and Pupils*, London, Stationery Office
Labour Party (1922) *Secondary Education for All: A Policy for Labour*, London, Labour Party
Report of the committee appointed by the Prime Minister to enquire into the position of classics in the educational system of the United Kingdom (1921)
Report of a Conference on secondary education in England, October 10–11(1893), Oxford, Clarendon Press
Report of the Royal Commission on secondary education (1895) (Bryce Report)
Report of the Schools Inquiry Commission (1868) (Taunton Report)

Secondary Sources

Abbiss, J. (1998) "The 'New Education' Fellowship in New Zealand: Its Activity and Influence in the 1930s and 1940s," *New Zealand Journal of Educational Studies*, 33/1, pp. 81–93.
Addison, P. (1994) "Oxford and the Second World War." In Brian Harrison (ed.), *The History of the University of Oxford*, vol. VIII, The Twentieth Century, Oxford, Clarendon Press, pp. 167–88.
Alcorn, N. (1999) *To the Fullest Extent of His Powers: C.E. Beeby's Life in Education*, Wellington, Victoria University Press.
Allsobrook, D. (1986) *Schools for the Shires: The Reform of Middle-Class Education in Mid-Victorian England*, Manchester, Manchester University Press.
Arnold, M. (1865/1964) *Schools and Universities on the Continent*, edited by R.H. Super, Ann Arbor, University of Michigan Press.
——— (1869/1932) *Culture and Anarchy*, edited by J. Dover Wilson, Cambridge, Cambridge University Press.
Aronowitz, S. (2003) *How Class Works: Power and Social Movements*, New Haven, Yale University Press.
Ball, S. (2003) *Class Strategies and the Education Market: The Middle Classes and Social Advantage*, London, Routledge Falmer.
Bamford, T.W. (1960) *Thomas Arnold*, London, The Cresset Press.
Banks, O. (1955) *Parity and Prestige in Secondary Education: A Study in Educational Sociology*, London, Routledge and Kegan Paul.
Barkan, E. (1992) *The Retreat of Scientific Racism: Changing Concepts of Race in Britain and the United States between the World Wars*, Cambridge, Cambridge University Press.
Betjeman, J. (1960) *Summoned by Bells*, London, John Murray.
Blades, B. (2003) "Deacon's School, Peterborough, 1902–1920: A Study of the Social and Economic Function of Secondary Schooling," PhD thesis, University of London.
Bledstein, B.J., and Johnston, R.D. (eds.) (2001) *The Middling Sorts: Explorations in the History of the American Middle Class*, London, Routledge.
Bourke, J. (2005) *Fear: A Cultural History*, London, Virago.
Bowen, J. (1989) "Education, Ideology and the Ruling Class: Hellenism and English Public Schools in the Nineteenth Century." In G.W. Clarke (ed.), *Rediscovering Hellenism: The Hellenic Inheritance and the English Imagination*, Cambridge, Cambridge University Press, pp. 161–86.
Bowra, C.M. (1966) *Memories, 1898–1939*, London, Weidenfeld and Nicolson.
Bradley, A.G., Champneys, A.C., and Baines, J.W. (1893) *A History of Marlborough College during Fifty Years from its Foundation to the Recent Time*, London, John Murray.
Brooks, R. (1993) "Lord Eustace Percy and the Abolition of the Compulsory, Elementary Curriculum in 1926," *Contemporary Record*, 7/1, pp. 86–102.
Brown, N., and Szeman, I. (eds.) (2000) *Pierre Bourdieu: Fieldwork in Culture*, Lanham, MD, Renman and Littlefield.
Butler, R.A. (1973) *The Art of the Possible*, London, Penguin.
Butterfield, H. (1931/1973) *The Whig Interpretation of History*, London, Penguin.
Campbell, A.E. (ed.) (1938) *Modern Trends in Education: The Proceedings of the New Education Fellowship Conference held in New Zealand in July 1937*, Wellington, Whitcombe and Tombs.
Card, T. (1994) *Eton Renewed: A History from 1860 to the Present Day*, London, John Murray.
Carter, M. (2003) *Anthony Blunt: His Lives*, London, Picador.
Clarke, F. (1940) *Education and Social Change: An English Interpretation*, London, Sheldon Books.

Coade, T.F. (ed.) (1931) *Harrow Lectures on Education*, Cambridge, Cambridge University Press.
Costello, J. (1988) *Mask of Treachery*, London, Collins.
Crossick, G. (1991) "From Gentlemen to the Residuum: Languages of Social Description in Victorian Britain." In Penelope Corfield (ed.), *Language, History and Class*, London, Basil Blackwell, pp. 150–178.
Davidoff, L., and Hall, C. (1987) *Family Fortunes: Men and Women of the English Middle Class, 1780–1950*, London, Hutchinson.
Devine, F. (2004) *Class Practices: How Parents Help Their Children to Get Good Jobs*, Cambridge, Cambridge University Press.
Devine, F., and Savage, M. (2005) "The Cultural Turn: Sociology and Class Analysis." In Fiona Devine, Mike Savage, John Scott, Rosemary Crompton (eds.), *Rethinking Class: Culture, Identities and Lifestyles*, London, Palgrave Macmillan, pp. 1–23.
Dickens, C. (1850/1948) *The Personal History of David Copperfield*, Oxford, Oxford University Press.
Draper, F.W.M. (1961) *Four Centuries of Merchant Taylors' School, 1561–1961*, London, Oxford University Press.
Durkheim, E. (1977) *The Evolution of Educational Thought: Lectures on the Formation and Development of Secondary Education in France*, London, Routledge and Kegan Paul.
Farr, W.C. (ed.) (1929) *Merchant Taylors' School: Its Origin, History and Present Surroundings*, London, Basil Blackwell.
Fletcher, F. (1937) *After Many Days: A Schoolmaster's Memories*, London, Robert Hale and Company.
Franklin, B.M., and McCulloch, G. (2003) "Partnerships in a 'Cold Climate': The Case of Britain." In B. Franklin, M. Bloch, T. Popkewitz (eds.), *Educational Partnerships and the State: The Paradoxes of Governing Schools, Children, and Families*, New York, Palgrave Macmillan.
Furlong, A. (2005) "Maintaining Middle Class Advantage," *British Journal of Sociology of Education*, 26/5, pp. 683–85.
Gathorne-Hardy, J. (1977) *The Public School Phenomenon, 597–1977*, London,
Gay, P. (1984) *The Bourgeois Experience, Victoria to Freud*, vol. I, Education of the Senses, Oxford, Oxford University Press.
Goodman, J., and Martin, J. (2004) *Women and Education, 1800–1980*, London, Palgrave Macmillan.
Graves, J. (1943) *Policy and Progress in Secondary Education, 1902–1942*, London, Thomas Nelson.
Grenfell, M. (1998) *Bourdieu and Education: Acts of Practical Theory*, London, Routledge Falmer.
Grimley, M. (2004) *Citizenship, Community and the Church of England: Liberal Anglican Theories of the State between the Wars*, Oxford, Clarendon Press.
Halsey, A.H. (1954) "The Relation between Education and Social Mobility with Reference to the Grammar School since 1944," PhD thesis, University of London.
Hearnshaw, L. (1979) *Cyril Burt, Psychologist*, London, Hodder and Stoughton.
Hill, C.P. (1951) *The History of Bristol Grammar School*, London, Pitman.
Hillier, B. (1988) *Young Betjeman*, London, John Murray.
Honey, J. (1977) *Tom Brown's Universe: The Development of the Public School in the 19th Century*, London, Millington Books.
Hughes, T. (1857/1949) *Tom Brown's Schooldays*, London, Dent.
Hunt, F. (1985) "Social Class and the Grading of Schools: Realities in Girls' Secondary Education, 1880–1940." In June Purvis (ed.), *The Education of Girls and Women*, Leicester, History of Education Society, pp. 27–46.
——— (1987) "Divided Aims: The Educational Implications of Opposing Ideologies in Girls' Secondary Schooling, 1850–1940." In F. Hunt (ed.), *Lessons for Life: The Schooling of Girls and Women, 1850–1950*, London, Basil Blackwell.

────── (1991) *Gender and Policy in Secondary Education: Schooling for Girls, 1902–44*, London, Harvester Wheatsheaf.
Kidd, A., and Nicholls, D. (eds.) (1998) *The Making of the British Middle Class? Studies of Regional and Cultural Diversity since the Eighteenth Century*, Stroud, Sutton Publishing.
────── (eds.) (1999) *Gender, Civic Culture and Consumerism: Middle-Class Identity in Britain, 1800–1940*, Manchester, Manchester University Press.
Laborde, E.D. (ed.) (1935) *Education of To-Day*, Cambridge, Cambridge University Press.
────── (ed.) (1939) *Problems in Modern Education*, Cambridge, Cambridge University Press.
Lawn, M. (1987) "The Spur and the Bridle: Changing the Mode of Curriculum Control," *Journal of Curriculum Studies*, 19/3, pp. 227–36.
Mangan, J.A. (1986) *Athleticism in the Victorian and Edwardian Public School*, London, Routledge Falmer.
Manton, K. (2001) *Socialism and Education in Britain, 1883–1902*, London, Woburn.
Marlborough College 1843–1943, Cambridge, Cambridge University Press.
Marsden, W.E. (1982) "Schools for the Urban Lower Middle Class: Third Grade or Higher Grade?" In Searby (ed.), *Educating the Victorian Middle Class*, pp. 45–85.
────── (1987) *Unequal Educational Provision in England and Wales: The Nineteenth Century Roots*, London, Woburn.
────── (1991) *Educating the Respectable: A Study of Fleet Road Board School, Hampstead, 1879–1903*, London, Woburn.
McCrum, M. (1989) *Thomas Arnold, Head Master: A Reassessment*, Oxford, Oxford University Press.
McCulloch, G. (1991) *Philosophers and Kings: Education for Leadership in Modern England*, Cambridge, Cambridge University Press.
────── (1993) "'Spens v. Norwood': Contesting the Educational State?" *History of Education*, 22/2, pp. 163–80.
────── (1994) *Educational Reconstruction: The 1944 Education Act and the Twenty-First Century*, London, Woburn.
────── (1998) *Failing the Ordinary Child? The Theory and Practice of Working Class Secondary Education*, Buckingham, Open University Press.
────── (2002) "Local Education Authorities and the Organisation of Secondary Education, 1943–1950," *Oxford Review of Education*, 28/2–3, pp. 235–46.
────── (2004a) "From Incorporation to Privatisation: Public and Private Secondary Education in Twentieth-Century England." In Richard Aldrich (ed.), *Public or Private Education?: Lessons from History*, London, Woburn, pp. 53–72.
────── (2004b) *Documentary Research in Education, History and the Socal Sciences*, London, Routledge Falmer.
────── (2006a) "Cyril Norwood and the English tradition of education," *Oxford Review of Education*, 32/1, pp. 55–69.
────── (2006b) "Education and the Middle Classes: The Case of the English Grammar Schools, 1868–1944," *History of Education*, 35/6, pp. 689–704.
McCulloch, G., and McCaig, C. (2002) "Reinventing the Past: The Case of the English Tradition of Education," *British Journal of Educational Studies*, 50/2, pp. 238–53.
McKibbin, R. (1998) *Classes and Cultures: England 1918–1951*, Oxford, Oxford University Press.
MacNeice, L. (1965) *The Strings are False: An Unfinished Autobiography*, London, Faber and Faber.
Mills, C. Wright (1959) *The Sociological Imagination*, London, Oxford University Press.
Morris, R.J. (2004) "A Year in the Public Life of the British Bourgeoisie." In Robert Colls, Richard Rodger (eds.), *Cities of Ideas: Civil Society and Urban Governance in Britain, 1800–2000*, Aldershot, Ashgate, pp. 121–43.
Nichols, J.G., and Lyons, P.A. (1876) *A History of the General Parish of Whalley*, 4th edition, vol. II, London.

Norwood, C. (1927) *The English Educational System*, London, Ernest Benn Ltd.
——— (1929) *The English Tradition of Education*, London, John Murray.
——— (1932) *Religion and Education*, London, Teaching Christ papers no IX, St Christopher Press, Society for Promoting Christian Knowledge.
——— (1937) *The Curriculum in Secondary Schools*, London, Association for Education in Citizenship.
Norwood, C., and Hope, A.H. (1909) *The Higher Education of Boys in England*, London, John Murray.
Norwood, S. (1876) *Our Indian Empire: The History of the Wonderful Rise of British Supremacy in Hindustan*, London, Samuel Tinsley.
Part, A. (1990) *The Making of a Mandarin*, London, Andre Deutsch.
Pedersen, J.S. (1987) *The Reform of Girls' Secondary and Higher Education in Victorian England: A Study of Elites and Educational Change*, London, Garland.
Playfair, G. (1937) *My Father's Son*, London, Geoffrey Bles.
Power, S., Edwards, T., Whitty, G., and Wigfall, V. (2003) *Education and the Middle Class*, Buckingham, Open University Press.
Reay, D. (2005) "Thinking Class, Making Class," *British Journal of Sociology of Education*, 26/1, pp. 139–43.
Reeder, D. (1987) "The Reconstruction of Secondary Education in England, 1869–1920." In D. Muller, F. Ringer, B. Simon (eds.), *The Rise of the Modern Educational System: Structural Change and Social Reproduction 1879–1920*, Cambridge, Cambridge University Press, pp. 135–50.
Reese, W.J. (2005) *America's Public Schools: From the Common School to "No Child Left Behind,"* Baltimore, Johns Hopkins University Press.
Roach, J. (1979) "Examinations and the secondary schools, 1900–1945," *History of Education*, 8/1, pp. 45–58.
——— (1986) *A History of Secondary Education in England, 1800–1870*, London, Longman.
——— (1991) *Secondary Education in England 1870–1902: Public Activity and Private Enterprise*, London, Routledge.
Russell, B. (1926) *On Education, Especially Early Childhood*, London, Allen and Unwin.
Searby, P. (ed.) (1982) *Educating the Victorian Middle Class*, Leicester, History of Education Society. History of Education Society.
Selby-Bigge, L. (1927) *The Board of Education*, London, Putnam's.
Selleck, R.J.W. (1994) *James Kay-Shuttleworth: Journey of an Outsider*, London, Woburn.
Simon, B. (1960) *The Two Nations and the Educational Structure, 1780–1870*, London, Lawrence and Wishart.
——— (1974) *The Politics of Educational Reform, 1920–1940*, London, Lawrence and Wishart.
——— (1986) "The 1944 Education Act: A Conservative Measure?" *History of Education*, 15/1, pp. 31–43.
——— (2002) "The 1902 Education Act—A Wrong Turning," *History of Education Society Bulletin*, 70, pp. 69–75.
——— (2004) "David Reeder's 'Alternative System': The School Boards in the 1890s." In Robert Colls, Richard Rodger (eds.), *Cities of Ideas: Civil Society and Urban Governance in Britain, 1800–2000*, Aldershot, Ashgate, pp. 178–206.
Simpson, D.J. (2005) *John Dewey and the Art of Teaching: Toward Reflective and Imaginative Practice*, London, Sage.
Skeggs, B. (2004) *Class, Self, Culture*, London, Routledge.
Smith, P., and Summerfield, G. (eds.) (1969) *Matthew Arnold and the Education of a New Order*, Cambridge, Cambridge University Press.

Stallworthy, J. (1995) *Louis MacNeice*, London, Faber and Faber.
Tawney, R.H. (1931/1964) *Equality*, edited by Richard Titmuss, London, Unwin Books.
Terrill, R. (1974) *R.H. Tawney and His Times: Socialism as Fellowship*, London, Deutsch.
Thompson, F.M.L. (1988) *The Rise of Respectable Society: A Social History of Victorian Britain, 1830–1900*, London, Fontana.
Tomlinson, S. (2005) *Education in a Post-Welfare Society*, 2nd edition, Maidenhead, Open University Press.
Trainor, R. (1991) "The Middle Class." In Martin Daunton (ed.), *The Cambridge Urban History of Britain*, vol. III, 1840–1950, Cambridge, Cambridge University Press, pp. 673–713.
Turner, F.M. (1989) "Why the Greeks and not the Romans in Victorian Britain?" In G.W. Clarke (ed.), *Rediscovering Hellenism: The Hellenic Inheritance and the English Imagination*, Cambridge, Cambridge University Press, pp. 61–81.
Turner, G.C. (1971) "Norwood, Sir Cyril (1875–1956)," *Dictionary of National Biography*, 1951–1960.
Tyerman, C. (2000) *A History of Harrow School*, London, Oxford University Press.
Vlaeminke, M. (2000) *The English Higher Grade Schools: A Lost Opportunity*, London, Woburn.
────── (2002) "Supreme Achievement or Disastrous Package? The 1902 Act revisited," *History of Education Society Bulletin*, 70, pp.76–87.
Walcott, F.G. (1970) *The Origins of Culture and Anarchy: Matthew Arnold and Popular Education in England*, Toronto, Toronto University Press.
Walford, G. (ed.) (2003) *British Private Schools: Research on Policy and Practice*, London, Woburn.
Walford, G., and Pickering, W. (eds.) (1998) *Durkheim and Modern Education*, London, Routledge.
West, F. (1984) *Gilbert Murray: A Life*, London, Croom Helm.
White, J. (1975) "The End of the Compulsory Curriculum." In Paul H. Hirst (ed.), *The Curriculum: The Doris Lee Lectures*, Studies in Education, New Series 2, Institute of Education, University of London, pp. 22–39.
Whitehead, A. (1932) *The Aims of Education and Other Essays*, London, Williams and Norgate.
Wilkinson, R. (1964) *The Prefects: British Leadership and the Public School Tradition*, London, Oxford University Press.
Wilson, J. Dover (ed.) (1928) *The Schools of England: A Study in Renaissance*, London, Sidgwick and Jackson.
Wolpe, A.M. (1974) "The Official Ideology of Education for Girls," in M. Flude, J. Ahier (eds.), *Educability, Schools and Ideology*, London, Croom Helm, pp. 138–59.
Worsley, T.C. (1940) *Barbarians and Philistines: Democracy and the Public Schools*, London, Robert Hale Ltd.
────── (1966) *Flannelled Fool: A Slice of Life in the Thirties*, London, Alan Ross Ltd.
Young, L. (2003) *Middle Class Culture in the Nineteenth Century: America, Australia and Britain*, London, Palgrave Macmillan.

Index

Admiralty, 2, 38
Advisory Council on Education, 152–53
Alington, Cyril, 90–91
Allied Schools, 117, 131, 132
Amis, Kingsley, 139–40
Aristotle, 21
Arnold, Matthew, vi, 11, 18–21, 25, 60, 105, 108–109, 112, 122, 149, 155, 156
Arnold, Thomas, vi, 21–23, 25, 58, 60, 79–80, 95, 99, 103, 104–105, 106, 108, 112, 122
Aronowitz, Stanley, 13
Association for Education in Citizenship, 118, 122, 124
Association of Education Committees, 143
Association of Head Mistresses, 69–70, 73–74, 101
athletics, 83, 110
Attlee, Clement, 152
Australia, 117, 128, 130

Bailey, Cyril, 118
Baker, William, 35, 36, 38
Ball, Stephen, 15
Banks, Olive, 5–6
Barbour-Simpson, Alexander, 131
Barker, Ernest, 121
Beeby, C.E., 129, 156
Betjeman, John, 83–84
Blades, Barry, 14
Bledstein, Burton J., 13
Blunt, Anthony, 84–85
Board of Education, 4, 6, 46, 48, 51, 58, 64, 72–76, 110, 132, 133, 142, 153
    consultative committee, 68–69, 72–73, 124, 144–45, 152

boarding schools, 16, 32, 109, 132–33, 142
Boer War, 41
Bourdieu, Pierre, 14
Bourke, Joanna, 13
Bowle, John, 86
Bradley, G.G., 80
Bristol, 7
    Grammar School, 2, 4, 7, 43–44, 45–61, 80, 81, 86, 140
    Young Men's Christian Association, 55
British Association, 76
British Empire, 11, 56, 107, 111, 120, 145
Brock, Dorothy, 141–42
Bruce, W.N., 64
Bryanston School, 118
Bryce, James, 16, 28
Bryce
    Commission, 15
    Report, 16–17
Burstall, Sara, 69
Burt, Cyril, 127, 156
Butler, Montagu, 94
Butler, R.A., 8, 142, 145, 148, 153
Butterfield, Herbert, 109

Caldicott, John, 45
Canada, ix, 98, 117, 118–19
Canford School, 131, 138
Canning, Clifford, 88, 118, 131
Carter, Miranda, 84
Cecil, Hugh, 133, 135
Charterhouse School, 80
Cheltenham College, 80
Church of England, 2, 79, 80, 93, 121–22, 131
Churchill, Winston, 142
civics curriculum, 54–55, 127

Civil Service, 2
Clarendon
  commission, 23, 35, 80
  Report, 34, 80
Clarke, Fred, 11–12, 127, 132, 152, 153
Clarke, M.G., 143
classics curriculum, 1, 22, 36, 37–38, 51, 65–66, 91, 93
Clifton College, 45, 46–47, 49, 54
Clive, Robert, 32
Coade, T.F., 113, 118
Cockerton judgement, 7
Colet, John, 58, 103
Common Entrance Examination, 103–104, 105
common schools, 145
continuation schools, 63
corporal punishment, 57
Cotton, G.E., 80
Council for Curriculum Reform, 150
cricket, 37, 40, 57, 85
Crossick, Geoffrey, 14
curriculum, 51–57, 64, 65–68, 82, 123, 124, 125–26, 128, 147, 150–51
  differentiation, 2, 4, 59, 68–72, 148–49
  see also individual subjects

Davidoff, Leonore, 14
day schools, 16, 27, 58, 109
De la Warr, Lord, 133–34
Dent, H.C., 115–16
Devine, Fiona, 14
Dewey, John, 156
Dickens, Charles, 24
Duckworth, F.R.G., 67, 132
Durkheim, Emile, vi, 1, 156, 157

Education Act
  (1902), 2, 4, 6–7, 11, 38, 42, 46, 59, 69, 74, 106, 115, 131
  (1918), 63, 72
  (1944), 1, 2, 4, 8, 10, 137, 152
education
  for citizenship, vi, 10, 55, 117–23, 125–26
  for girls, 4, 17, 61, 68–72, 107, 108
elementary education, 30, 34, 104, 109–10

Elementary Education Act (1870), 6, 7, 21, 28
Endowed Schools Commission, 28, 34
English curriculum, 46, 53–54, 124, 125–26, 128, 150–51
English Association, 150–51
English tradition, 2, 4, 58–59, 99–116, 134
Ensor, Beatrice, 128
Eton College, 13, 18–19, 31, 90, 123, 133
eugenics, 122
Europe, 18
examinations, 8, 51–52, 64, 67, 110, 141, 147, 149–50, 151
examination boards, 151

Farrar, F.W., 22
fascism, 117, 119, 128, 134–35
feminism, 6
Fiji, 129
Fisher, Herbert, 63–64, 153
Fletcher, Frank, 80
Ford, Lionel, 90, 94
France, 18, 58, 59, 60
Fraser, Peter, 129

Gathorne-Hardy, Jonathan, 90
Gay, Peter, 13–14
General Certificate of Education, 151
General Strike, 104
Germany, 58, 59, 60, 84, 85, 133, 134
Gibson, A.M., 126
Goodenough, Francis, 118
Gorell, Lord, 113
Gould, Frederick, 7
grammar schools, 2, 5, 13, 23, 25, 27–34, 73, 124, 143–44, 148–49, 152
Graves, John, 4–5
Greek curriculum, 36, 42, 53, 66
Green, T.H., 122
Greene, C.H., 64
Grimley, Matthew, 121
Gurner, Ronald, 113

Hadow, Henry, 73
Hadow Report, 72–73, 75, 76, 101–102, 140, 149
Hall, Catherine, 14
Halsey, A.H., 14

Index

Hamley, H.R., 127
Harrogate College, 131
Harrow School, 3, 4, 7, 19, 22, 35, 79, 88–99, 102, 117, 118, 119–20, 125, 137, 157
Hastings, O.M., 143
headmasters, 59–60, 94–95
Headmasters' Conference, 23, 109
Hessey, James, 35
high schools, 10
higher grade schools, 7, 21, 23, 45
Hill, A.V., 91–92
Hillard, Frederick, 47
history curriculum, 54–55
Holmes, Maurice, 74, 75, 144, 145, 153
Honey, John, 23, 99
Hope, Arthur H., 45, 58–61
Hughes, Thomas, 22, 25
Hunt, Felicity, 6, 14
Hutchings, A.W.S., 143

Incorporated Association of Head Masters (IAHM), 148
independent schools, vi, 2, 4, 11
 see also public schools
India, 32–33, 102
Inge, W.R., 113
Innes, P.D., 143
Institute of Education London, 127, 132
Irwin, Lord, 123
Italy, 111, 120

Jones, Joseph, 143
Jowett, Benjamin, 80

Kandel, Isaac, 127, 130
Kay-Shuttleworth, James, 3
Keynes, Lord, 153
Kidd, Alan, 14
Kilner, Walter, 40

labour
 market, 7
 movement, 7
Labour Party, 63, 72, 119
Larkin, Philip, 139–40
Latin curriculum, 31, 36, 52–53, 67, 91
Leeds, 39

Boys' Modern School, 113
Grammar School, 2, 4, 38–44, 80, 86, 143, 148
Leeson, Spencer, 118, 125
Leighton, Robert, 45, 46–48
Lindsay, A.D., 121
Liverpool Collegiate High School, 126
Livingstone, R.W., 66, 118
local education authorities, 64, 67, 107, 110, 133, 150, 152
Lowerison, Harry, 7
Lyon, Hugh, 126
Lyon, John, 91
Lyttleton, Edward, 123

McKibbin, Ross, 14
MacNeice, Louis, 84
Malim, F.B., 137
Manchester High School for Girls, 69, 143
Manton, Kevin, 7, 14
Marlborough College, 3, 4, 7, 22, 63, 79–88, 102, 134, 142, 147, 157
Marsden, W.E., 12, 23
Martin, Kingsley, 116
Marxism, 6, 12
Mary Datchelors School, 141
Matthews, J.H.D., 40
Matthias, Frederick J., 113–14
Merchant Taylors School, 2, 27, 34–38, 42, 44, 51
middle class, 3, 11–15, 16–17, 19–20, 111–12, 128, 145, 152, 155, 156–57
 education, 11–25, 38, 42–44, 46–47, 103
Mills, C. Wright, 3, 14, 102
Modern Churchmen's Union, 138
modern languages curriculum, 65
modern schools, 73, 140, 142
Morton, G.F., 113
Motion, Andrew, 139
Mulcaster, Richard, 35
multilateral schools, 145
Murray, Gilbert, 36, 38, 66
Murray, John, 113
Myers, J.E., 143

Naisbitt, E.W., 143
National Union of Teachers, 113, 125
Nettleship, R.L., 80

new education, 10
New Education Fellowship, 117, 122, 128–30
New Zealand, 117, 128, 129–30
Nicholls, David, 14
Norwood, Barbara, 131
Norwood, Catherine, 3, 39–40, 89–90, 98, 118, 128–30
Norwood, Elizabeth, 33, 40
Norwood, Enid, vi, 43, 88
Norwood, Samuel, 2, 27–34, 38, 41, 44, 58, 99, 102
Norwood
 Committee, 141–50
 Report, vi, 2, 8–10, 147–52, 157
Notting Hill and Bayswater High School, 39

Oxford Society, 138

parity of esteem, 140–41
Part, Anthony, 91
Pelham, E.H., 66, 69–70
Pember, Francis, 92, 96, 98
Percy, Lord Eustace, 65
Petch, J.A., 151
Plato, vi, 21, 63, 112, 122, 134, 138, 148, 149
Playfair, Giles, 89
Pope, C.G., 95–99, 105, 113
Power, Sally, 15
preparatory schools, 91, 105, 145
Preston, George, 28
primary schools, 145
psychology, 127, 149
public schools, 2, 4, 16, 19, 21–23, 30, 37, 51, 102–107, 131–35, 142, 149, 155

Reeder, David, 12, 23
religion, 22, 39, 64, 108, 120–21, 138, 144
Richardson, W.R., 75
Richmond Grammar School, 96
Roach, John, 17
Rothschild, Victor, 91, 95–96
rugby football, 37, 57, 85, 95, 97
Rugby School, 19, 21–23, 58, 80, 126
Russell, Bertrand, 111

Sadler, Michael, 115, 156

St Paul's School, 58, 103
Salisbury, bishop of, 80
Savage, Mike, 14
scholarships, 49–50
school
 boards, 7, 23, 28
 cadet corps, 55–56
 debating societies, 37, 40–41, 57
 discipline, 48–49
 endowments, 28
 fees, 35, 49–50, 91
 inspections, 46, 88, 91, 92–93
 library, 57
 selection, 2
 tradition, 82, 94–99, 101–16
School Certificate examination, 73–74, 75, 77, 86, 104, 124, 140, 147, 149
science curriculum, 56, 65, 82, 91–92, 122, 123, 128
Science Masters' Association, 122
scouting, 55
Seaford School, 131
Searby, Peter, 12
Secondary School Regulations, 51, 69
Secondary Schools Examinations Council (SSEC), 3, 8, 63, 64, 73, 101, 141, 151, 152
security, 11–12
Selby-Bigge, Lewis, 106
Selleck, Richard, 3
Shaftesbury, Lord, 121
Sharp, Percival, 143, 145
Sharrock, S.H., 143
Sidgwick, Arthur, 36
Simon, Brian, 5–6, 7–8, 23, 155
Simon, Ernest, 118, 124–25
Simon, Shena, 124, 140
Skeggs, Beverley, 14
Smith, George, 64
social class, 5, 61, 87–88, 103, 105, 149, 156
Soviet Union, 12, 111
Spence, C.H., 54
Spens, Will, 124, 142, 144–45, 146
Spens
 Committee, 123–28, 145
 Report, 126–28, 140, 141, 143, 144, 145, 149

sports, 23, 57, 83, 84, 95, 138
Stead, F.B., 67–68, 70, 74, 75, 93, 123
Storr, Francis, 35–36
Stowe School, 131
Taunton
  Commission, 11, 15, 28–29, 33, 58, 60
  Report, 15–16, 17, 18, 149, 155
Tawney, R.H., 6, 12–13, 72, 119, 156
teachers, vi, 52, 64, 67–68, 107, 110, 146–47, 149
technical high schools, 140, 141, 142
Temple, William, 121
Thomas, Terry, 143, 148
Thompson, F.M.L., 14
Thorne, Robert, 45
Tomlinson, Sally, 15
Toulouse Lyceum, 19
tradition, 9, 32–33, 44, 63, 71–72, 75, 77, 101–16, 143
Trainor, Richard, 14
Turner, George, 4, 147
Tyerman, Christopher, 89

United States of America, 10, 13, 58, 59, 117, 120–21, 128, 129, 145
University of Oxford, 2, 36, 138–39
  Balliol College, 50, 80, 121
  conference on secondary education, 16
  Department of Education, 138
  Corpus Christi, 21
  Oriel College, 21
  Queen's College, 42–43
  St John's College, 2, 35, 38, 50, 89, 137–40, 147
University of Sheffield, 4
  Norwood archive, 4

Valentine, C.W., 113
Vaughan, C.J., 22, 91
Vlaeminke, Mel, 6–7, 14, 23

Walcott, Fred G., 25
Warrington, P.E., 131
welfare, 13
Wellington College, 80, 137
Wells, H.G., 41
Westonbirt School, 71, 131
Whalley, 32
  Abbey, 27–28
  Grammar School, 27–34, 99
White, Thomas, 34
White Paper on education
  (1943), 8, 150
  (2005), 157
Whitehead, Alfred, 66, 123
Whitgift Grammar School, 111
Wilkinson, Ellen, 153
William of Wykeham, 58, 103, 104, 108
Williams, G.G., 141, 143, 146
Williams, W. Nalder, 143
Winchester College, 19, 58, 103, 104, 108
Wood, Robert, 140–41
World War
  I, 56, 61, 63, 81–82, 107, 122
  II, 2, 55, 84, 117, 122, 132, 139, 141, 157
Worsley, T.C., 83, 88, 134–35
Wrekin College, 131
Wynne-Edwards, J.R., 40

Young, R.F., 126

GPSR Compliance
The European Union's (EU) General Product Safety Regulation (GPSR) is a set of rules that requires consumer products to be safe and our obligations to ensure this.

If you have any concerns about our products, you can contact us on

ProductSafety@springernature.com

In case Publisher is established outside the EU, the EU authorized representative is:

Springer Nature Customer Service Center GmbH
Europaplatz 3
69115 Heidelberg, Germany

www.ingramcontent.com/pod-product-compliance
Lightning Source LLC
LaVergne TN
LVHW012101070526
838200LV00074BA/3894